# Università degli Studi di Napoli "Federico II"

Scuola Politecnica e delle Scienze di Base

Dipartimento di Ingegneria Industriale

Corso di Dottorato in Ingegneria Industriale

XXXI Ciclo

Tesi di Dottorato

# Experimental and numerical investigation of Rayleigh-Bénard convection

## Gerardo Paolillo

**Tutori:**
Prof. Ing. Gennaro Cardone
Dott. Ing. Carlo Salvatore Greco

**Coordinatore:**
Prof. Ing. Michele Grassi

Dicembre 2018

Titolo | Experimental and numerical investigation
         of Rayleigh-Bénard convection
Autore | Gerardo Paolillo
ISBN | 978-88-31607-77-3



Youcanprint
Via Marco Biagi 6 - 73100 Lecce
www.youcanprint.it
info@youcanprint.it

*Your assumptions are*
*your windows on the world.*
*Scrub them off every once in a while,*
*or the light won't come in.*

Isaac Asimov

# Abstract

This thesis presents a combined experimental and numerical investigation of Rayleigh-Bénard convection without and with a background rotation. Rayleigh-Bénard convection, the buoyancy-driven flow induced by temperature gradients, is relevant to a wide variety of both natural phenomena (of which motions in the atmosphere and in the oceans are only the most straightforward examples) and technological applications (like melting of pure metals or flows in turbomachinery). Despite the numerous works on the subject, different aspects of this phenomenon are still unclear or deserving of further investigation. Due to its inherently turbulent nature, the analysis of the flow field makes three-dimensional measurements mandatory to get an unambiguous picture of the underlying rich dynamics. Three-dimensional whole-field velocity measurements are very rare in literature and anyway very recent; in consideration of this, the present work focuses on the application of a state-of-art optical investigation technique, namely the tomographic particle image velocimetry, to the study of such a phenomenon. An experimental apparatus suitable for this purpose is designed and developed. Some technical issues inherent to the optical measurements are extensively addressed; in particular, an innovative camera model is formulated to precisely account for the optical distortions caused by the cylinder sidewall, through which the measurement volume is imaged. In addition to experimental measurements, direct numerical simulations are performed, in which the non-adiabaticity of the lateral wall is accounted for by simulating the presence of the wall itself with its physical properties. The comparison between the experimental and numerical results offers the chance of validating the physical models and computational approaches used in the numerical environment and, at the same time, pointing out unavoidable non-idealities of the experimental setup that make the phenomenon to differ from the canonical problem addressed not only numerically, but also theoretically.

# Contents

# List of symbols

## Acronyms

| | |
|---|---|
| 2D | two-dimensional |
| 3D | three-dimensional |
| ART | algebraic reconstruction technique |
| BM | Bénard-Marangoni |
| CCD | charge-coupled device |
| CMOS | complementary metal-oxide-semiconductor |
| CSMART | camera sequential multiplicative algebraic reconstruction technique |
| DLT | direct linear transformation |
| DNS | direct numerical simulation |
| DRS | double-roll state |
| FFT | fast Fourier transform |
| GL | Grossman-Lohse |
| IPR | iterative particle reconstruction |
| LED | light emitting diode |
| LOR | low-order reconstruction |
| LOS | line of sight |
| LSC | large-scale circulation |
| MART | multiplicative algebraic reconstruction technique |
| MGIWD | multi-grid iterative window deformation |
| MTE | motion tracking-enhanced |
| Nd:YAG | Neodymium Yttrium Aluminum Garnet |
| Nd:YLF | Neodymium Yttrium Lithium Fluoride |
| OB | Oberbeck-Boussinesq |
| PC | personal computer |
| PID | proportional integral derivative |
| PIV | particle image velocimetry |
| PMMA | polymethyl methacrylate |
| POD | proper orthogonal decomposition |
| PTV | particle tracking velocimetry |
| RB | Rayleigh-Bénard |
| RTD | resistance temperature detector |
| sCMOS | scientific complementary metal-oxide-semiconductor |
| SMART | sequential multiplicative algebraic reconstruction technique |
| SMTE | sequential motion tracking-enhanced |
| SR | single roll |
| SRS | single-roll state |

| STB | shake-the-box |
| SVD | singular-value decomposition |
| TEC | thermo-electric controller |
| T-PIV | tomographic particle image velocimetry |
| TS | transitional state |

## Roman letters

| | |
|---|---|
| $A$ | counterpart of the generic object point on the undistorted ray |
| $\mathbf{A}, \mathbf{B}$ | observation matrices |
| $A_b, A_m, A_t$ | amplitude of the cosine fit of the azimuthal vertical velocity profile at bottom ($x/L = 0.25$), middle ($x/L = 0.5$) and top ($x/L = 0.75$) of the convection cell |
| $\mathbf{a}_j, \mathbf{b}_j$ | observations |
| $A_p$ | particle image area, $\text{pixel}^{-2}$ |
| $A_{u_j}, A_j$ | amplitude of the cosine fit of the azimuthal vertical velocity profile at the $j$-th height |
| $A_{\theta_j}$ | amplitude of the cosine fit of the azimuthal temperature profile at the $j$-th height |
| $\mathbf{B}$ | linear distortion matrix |
| $b_1, b_2$ | linear distortion coefficients |
| $\mathbf{B}^{(C)}, \mathbf{B}^{(D)}$ | correlated and decorrelated part of observation matrix |
| $\mathbf{b}_j^{(C)}, \mathbf{b}_j^{(D)}$ | correlated and decorrelated part of observations |
| $C$ | center of projection |
| $\mathcal{C}$ | cross-correlation function |
| $C'$ | principal point |
| $c$ | index of camera |
| $C_{pf}$ | fluid specific heat, $\text{J}/(\text{kg K})$ |
| $C_{ps}$ | sidewall specific heat, $\text{J}/(\text{kg K})$ |
| $\mathbf{C_V}$ | temporal covariance matrix |
| $\mathbf{C'_V}$ | spatial covariance matrix |
| $D$ | cylinder diameter, m |
| $d_\text{diff}$ | diffraction particle image diameter, m |
| $d_\text{geom}$ | geometric particle image diameter, m |
| $d_p$ | particle diameter, m |
| $d_\text{pix}$ | pixel size, m |
| $d_\tau$ | particle image diameter, m |
| $d_\tau^*$ | particle image diameter in pixel units |
| $d_X, d_Y$ | components of the disparity vector, pixel |
| $E$ | light intensity distribution |
| $\mathbf{E}$ | vector of voxel light intensities |
| $E^{(ij)}$ | kinetic energy contribution of the $i$-th and $j$-th POD modes |
| $E_\text{max}^{(ij)}$ | time maximum of the kinetic energy contribution of the $i$-th and $j$-th POD modes |

| $\mathbf{F}$ | additional resultant particle acceleration, $\mathrm{m/s^2}$ |
| $\mathbf{f}$ | non-dimensional additional resultant particle acceleration |
| $f_{\#}$ | $f$-number |
| $G$ | origin of the cylinder reference frame |
| $g$ | gravity acceleration, $\mathrm{m/s^2}$ |
| $\hat{\mathbf{g}}$ | gravity unit vector |
| $I$ | recorded light intensity |
| $\mathbf{I}$ | vector of pixel light intensities |
| $i$ | index of pixel |
| $\mathbf{I}_m, \mathbf{I}_n$ | unity matrices of order $m, n$ |
| $j$ | index of voxel |
| $\mathbf{K}$ | perspective projection matrix |
| $k$ | index of iteration |
| $k_1, k_2$ | radial distortion coefficients |
| $L$ | height of the fluid layer/reference length, m |
| $l$ | number of pixels |
| $M$ | magnification factor |
| $m$ | number of voxels or size of observation |
| $\mathbf{m}_c$ | mapping function |
| $N$ | number of cameras |
| $n$ | index for time instant or number of observations |
| $N_g$ | number of ghost particles |
| $n_i$ | refractive index of the medium on the side of the incident ray |
| $N_p$ | number of actual particles |
| $n_r$ | refractive index of the medium on the side of the refractive ray |
| $N_S$ | source density |
| $N_w$ | linear size of the interrogation volume |
| $N_x, N_y, N_z$ | orders of the polynomial mapping functions |
| $O$ | origin of the world reference frame |
| $O'$ | origin of the image reference frame |
| $P$ | reduced pressure, Pa, or object point |
| $p$ | non-dimensional reduced pressure |
| $p_1, p_2$ | decentering distortion coefficients |
| $P'$ | image point |
| $\mathbf{p}_c$ | calibration parameters |
| $ppp$ | particle per pixel, $\mathrm{pixel^{-2}}$ |
| $P_s$ | static pressure, Pa |
| $Q$ | second invariant of velocity gradient tensor, $\mathrm{s^{-2}}$, or projection of $P$ on the principal axis |
| $\dot{q}$ | total heat flux, $\mathrm{W/m^2}$ |
| $\dot{q}_s$ | local heat flux through the cylinder sidewall, $\mathrm{W/m^2}$ |
| $\mathbf{R}$ | radius vector, m, or rotation matrix |
| $r$ | non-dimensional radial coordinate |
| $\mathbf{r}$ | non-dimensional radius vector |
| $\hat{\mathbf{r}}$ | radial unit vector |

| | |
|---|---|
| $R_f$ | fluid thermal resistance, $\mathrm{m^2K/W}$ |
| $r_i$ | cylinder internal radius, m |
| $R_p$ | plate thermal resistance, $\mathrm{m^2K/W}$ |
| $\mathbf{s}$ | displacement vector, m |
| $s_1, s_2$ | thin prism distortion coefficients |
| $S_b, S_m, S_t$ | normalized energy in the first Fourier of the azimuthal vertical velocity profile at bottom ($x/L = 0.25$), middle ($x/L = 0.5$) and top ($x/L = 0.75$) of the convection cell |
| $T$ | temperature, °C |
| $\mathbf{T}$ | world-to-camera reference frame transformation matrix |
| $t$ | time, s |
| $\mathbf{t}$ | origin translation between world and camera reference frames, m |
| $T_0$ | reference temperature, °C |
| $T_t$ | temperature of the top plate, °C |
| $T_e$ | uniform temperature at the external side of the cylinder sidewall, °C |
| $T_b$ | temperature of the bottom plate, °C |
| $T_m$ | average temperature over the fluid sample, °C |
| $t_p$ | plate thickness, m |
| $t_s$ | sidewall thickness, m |
| $t_w$ | wall thickness, m |
| $t_x, t_y, t_z$ | origin translations along coordinate axes between world and camera reference frames, m |
| $U$ | characteristic velocity scale, m/s |
| $\mathbf{U}$ | velocity vector, m/s |
| $u, v, w$ | non-dimensional velocity components in the Cartesian reference frame |
| $\mathbf{u}$ | non-dimensional velocity vector |
| $u'$ | temporal fluctuation of the non-dimensional vertical velocity |
| $u_0$ | free-fall velocity, m/s |
| $\mathbf{U}_p$ | particle velocity vector, m/s |
| $\mathbf{u}_p$ | non-dimensional particle velocity vector |
| $u_x, u_r, u_\varphi$ | non-dimensional velocity components in the cylindrical reference frame |
| $u'_x, u'_r, u'_\varphi$ | components of the non-dimensional velocity fluctuation in the cylindrical reference frame |
| $\mathbf{u}_\varphi$ | non-dimensional azimuthal velocity component vector |
| $\mathbf{V}$ | observation matrix |
| $v_{ij}$ | component of observation |
| $\mathbf{v}_j$ | observation |
| $W$ | recorded light distribution in the interrogation volume |
| $\mathbf{W}$ | matrix of weighting coefficients for tomographic reconstruction |
| $w_{ij}^{(c)}$ | weighting coefficient for tomographic reconstruction |
| $X, Y$ | coordinates in the image reference frame, pixel |
| $x, y, z$ | coordinates in the world reference frame, m |
| $\mathbf{x}$ | position vector, m |
| $X_0, Y_0$ | image coordinates of the principal point, pixel |
| $\mathbf{X}_c$ | image position vector, m |

$x_c, y_c, z_c$     coordinates in the camera reference frame, m
$X_p$     ratio of equivalent fluid thermal resistance to plate thermal resistance
$X_s$     ratio of equivalent fluid thermal resistance to sidewall thermal resistance

## Greek letters

| | |
|---|---|
| $\alpha$ | thermal diffusivity, $\mathrm{m^2/s}$ |
| $\alpha_c, \beta_c, \gamma_c$ | Euler angles of the cylinder reference frame, $^\circ$ |
| $\beta$ | thermal expansion coefficient, $\mathrm{^\circ C^{-1}}$ |
| $\beta_{jk}$ | POD coefficients |
| $\beta_{jk}^{(\mathrm{ext})}$ | extended POD coefficients |
| $\gamma_{jk}$ | coefficients of generic linear decomposition |
| $\Delta$ | temperature difference, K |
| $\delta_E$ | thickness of the Ekman boundary layer, m |
| $\Delta_{\mathrm{field}}$ | depth of field, m |
| $\delta_{lm}$ | Kronecker delta |
| $\varepsilon_s$ | search radius in self-calibration process, pixel |
| $\varepsilon_\theta$ | volume- and time-averaged heat dissipation rate, $\mathrm{J/(kg\,s)}$ |
| $\varepsilon_u$ | volume- and time-averaged kinetic energy dissipation rate, $\mathrm{J/(kg\,s)}$ |
| $\eta_\nu$ | Kolmogorov length scale, m |
| $\Theta$ | relative temperature, $\mathrm{^\circ C}$ |
| $\theta$ | non-dimensional relative temperature |
| $\theta_e$ | non-dimensional relative temperature at the external side of the cylinder sidewall |
| $\theta_j$ | average of the cosine fit of the azimuthal temperature profile at the $j$-th height |
| $\vartheta_i$ | angle between the incident ray and the normal of the surface, $^\circ$ |
| $\vartheta_r$ | angle between the refracted ray and the normal of the surface, $^\circ$ |
| $\kappa$ | thermal conductivity, $\mathrm{W/(mK)}$ |
| $\kappa_p$ | plate thermal conductivity, $\mathrm{W/(mK)}$ |
| $\kappa_s$ | sidewall thermal conductivity, $\mathrm{W/(mK)}$ |
| $\lambda$ | light wavelength, m, or Lagrange multiplier |
| $\mu$ | dynamic viscosity, $\mathrm{Pa\,s}$ |
| $\bar{\mu}$ | relaxation coefficient |
| $\mu_n$ | averaged recorded light intensity over the interrogation volume |
| $\nu$ | kinematic viscosity, $\mathrm{m^2/s}$ |
| $\xi, \eta, \zeta$ | coordinates in the reference frame attached to the cylinder, m |
| $\xi_0, \eta_0, \zeta_0$ | translations of the origin of the cylinder reference frame, m |
| $\rho$ | density, $\mathrm{kg/m^3}$ |
| $\rho_0$ | reference density, $\mathrm{kg/m^3}$ |
| $\rho_f$ | fluid density, $\mathrm{kg/m^3}$ |
| $\rho_p$ | particle density, $\mathrm{kg/m^3}$ |
| $\rho_s$ | sidewall density, $\mathrm{kg/m^3}$ |
| $\varrho$ | ratio of the refractive indexes |

| $\varrho_e$ | ratio of the refractive index of the external medium to the refractive index of the cylinder material |
| $\varrho_i$ | ratio of the refractive index of the cylinder material to the refractive index of the internal medium |
| $\mathbf{\Sigma}$ | matrix of singular values |
| $\sigma$ | singular value |
| $\mathbf{\Sigma}_{\mathbf{B}}^{(ext)}$ | matrix of extended singular values |
| $\tau_E$ | Ekman time scale, s |
| $\tau_f$ | characteristic time scale, s |
| $\tau_\nu$ | Kolmogorov time scale, s |
| $\tau_p$ | particle relaxation time, s |
| $\mathbf{\Phi}$ | matrix of right-singular vectors |
| $\Phi$ | LSC orientation identified by means of POD modes, $^\circ$ |
| $\phi, \delta, \psi$ | Euler angles of the camera reference frame, $^\circ$ |
| $\boldsymbol{\phi}$ | right-singular vector |
| $\phi_W$ | weighting function |
| $\varphi$ | azimuthal coordinate in cylindrical reference frame, $^\circ$ |
| $\varphi_b, \varphi_m, \varphi_t$ | initial phase of the cosine fit of the azimuthal vertical velocity profile at bottom ($x/L = 0.25$), middle ($x/L = 0.5$) and top ($x/L = 0.75$) of the convection cell, $^\circ$ |
| $\boldsymbol{\varphi}_k$ | modes of generic linear decomposition |
| $\varphi_{\theta_j}$ | initial phase of the cosine fit of the azimuthal temperature profile at the $j$-th height, $^\circ$ |
| $\varphi_{u_j}$ | initial phase of the cosine fit of the azimuthal vertical velocity profile at the $j$-th height, $^\circ$ |
| $\chi$ | pixel aspect ratio |
| $\mathbf{\Psi}$ | matrix of left-singular vectors |
| $\psi$ | potential for gravitational field, $\mathrm{m}^2/\mathrm{s}^2$ |
| $\boldsymbol{\psi}$ | POD mode or left-singular vector |
| $\mathbf{\Psi}_{\mathbf{B}}^{(ext)}$ | matrix of normalized extended POD modes |
| $\boldsymbol{\psi}_{\mathbf{B}}^{(ext)}$ | normalized extended POD modes |
| $\overline{\mathbf{\Psi}}_{\mathbf{B}}^{(ext)}$ | matrix of non-normalized extended POD modes |
| $\mathbf{\Omega}$ | angular velocity vector, $\mathrm{rad/s}$ |
| $\widehat{\mathbf{\Omega}}$ | angular velocity unit vector |
| $\omega_b$ | $x$-component of the vorticity at the edge of the boundary layer, $\mathrm{s}^{-1}$ |

## Other symbols

| $\mathcal{F}$ | Lagrange functional |
| $\delta_c P$ | ray distortion for the generic object point, pixel |
| $\Delta r$ | cylinder sidewall thickness, m |
| $\Delta U$ | characteristic slip velocity of the seeding particle, m/s |
| $\delta X, \delta Y$ | total distortions of the pinhole camera model, pixel |
| $\Delta Y$ | local projection error, pixel |

$\delta\varphi_{ij}$      difference between initial phase angles of the cosine fit of the azimuthal vertical velocity profile at the $i$-th and $j$-th heights, $^\circ$

## Operators

$\partial_t, \mathrm{d}_t$      time derivative, $\mathrm{s}^{-1}$
$\partial_x, \partial_y, \partial_z$      spatial derivative, $\mathrm{m}^{-1}$
$\nabla$      gradient, $\mathrm{m}^{-1}$
$\nabla^2$      Laplacian, $\mathrm{m}^{-2}$
$\langle\cdot\rangle_{(\cdot)}$      average along specified spatial directions
$\overline{(\cdot)}$      temporal average
$(\cdot,\cdot)$      scalar product between vectors
$(\cdot)^{\mathrm{T}}$      transpose of matrix
$(\cdot)^*$      conjugate transpose of matrix
$(\cdot)^+$      pseudo-inverse of matrix

## Dimensionless parameters

$Ar_p$      particle Archimedes number, $g\,d_p^3\,\rho\,(\rho_p - \rho)/(18\mu^2)$
$Ar_p'$      particle buoyancy Archimedes number, $g\,d_p^3\,\rho_0^2\,\beta\theta\Delta/(18\mu^2)$
$Ek$      Ekman number, $\nu/(\Omega L^2)$
$Fr$      Froude number, $\Omega^2 D/(2g)$
$Nu$      Nusselt number, $\dot{q}L/(\kappa\Delta)$
$Nu_\infty$      Nusselt number for adiabatic wall case, $\dot{q}_\infty L/(\kappa\Delta)$
$Nu_l$      Nusselt number through lateral wall, $\dot{q}_l L/(\kappa\Delta)$
$Pr$      Prandtl number, $\nu/\alpha$
$Ra$      Rayleigh number, $g\beta\Delta L^3/(\alpha\nu)$
$Re$      Reynolds number, $UL/\nu$
$Re_p$      particle Reynolds number, $\rho U d_p/\mu$
$Re_p'$      particle Reynolds number based on slip velocity, $\rho\Delta U d_p/\mu$
$Ro$      Rossby number, $\sqrt{\beta\Delta g/L}/(2\Omega)$
$Ro_c$      critical Rossby number, $\sqrt{\beta\Delta g/L}/(2\Omega)$
$St_p$      particle Stokes number, $\tau_p U/L$
$Ta$      Taylor number, $(2\Omega L^2/\nu)^2$
$\Gamma$      aspect ratio, $D/L$

# 1

# Introduction

## 1.1. Historical background and applications

THERMALLY-DRIVEN convection, i.e. the fluid motion induced by the density differences due to temperature gradients, is a transport process relevant to a wide variety of natural phenomena and technological applications. Such a process is characterized by the formation of patterns of motion and coherent structures, which were first meticolously investigated by Henri Bénard in 1900 [1, 2], although earlier evidences of the same date back half a century to the works of Weber [3] and Thompson [4].

Bénard used optical methods to analyze the behavior of very thin layers of different fluids standing on a copper plate maintained at a constant temperature with an upper surface in contact with ambient air. He observed that, by increasing the temperature of the lower surface beyond a certain threshold, the layer assumed a reticulated structure (figure 1.1a). Bénard referred to the constituent hexagonal cells of this structure as "cellular vortices" (*tourbillons cellulaires*) and described the inherent fluid motion, consisting of ascending hot currents at the center and descending cold currents at the periphery. Bénard's «careful and skillful experiments» inspired the theoretical work of Lord Rayleigh in 1916 [5]. By means of linear stability analysis, Rayleigh formulated a criterion for the onset of the convective motion in a layer of fluid, heated from below and delimited by two no-shear boundaries, in terms of the non-dimensional difference of temperature across the layer, the later defined Rayleigh number. Successive studies of linear stability were those of Jeffreys [6, 7] and Low [8]. In those years, Bénard noticed a satisfactory agreement between his measurements of the size of the convective cells and the predictions of the Rayleigh's theory [9]. However, a later analysis of the same results [10] showed a net discrepancy in terms of the critical value of the Rayleigh number (up to three orders of magnitude). The source of "the Rayleigh's deficiency" — as Bénard called such an inconsistency — was clarified only after two decades, when Block, in a

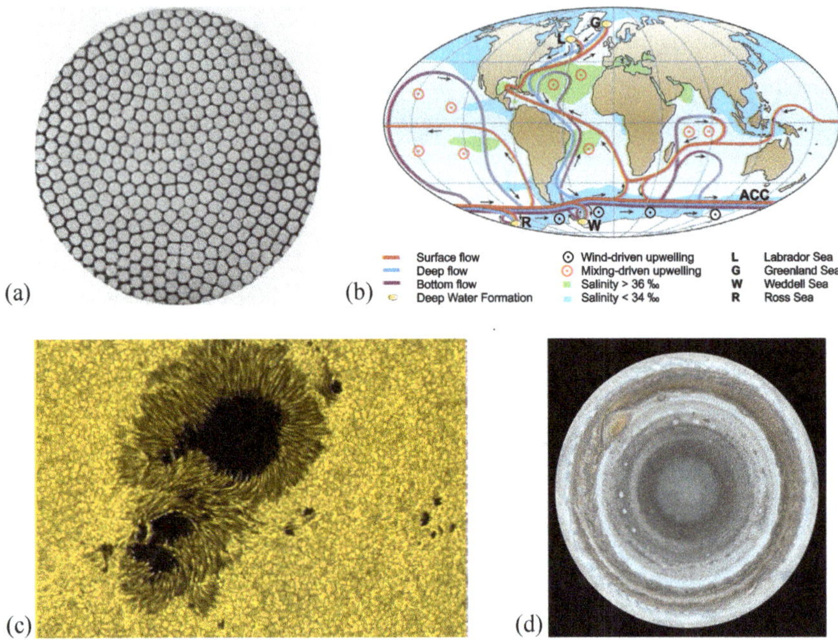

Figure 1.1: Some examples of Rayleigh-Bénard convection in nature. (a) Picture from the famous Bénard's experiment [13]: hexagonal cells in a layer of spermaceti heated from below. (b) Schematic representation of the global thermohaline circulation (source: Rahmstorf [14]). (c) Sunspot and solar granules imaged by Hinode's Solar Optical Telescope (source: Hinode JAXA/NASA [15]). (d) Polar map of Jupiter's South constructed from images taken by Cassini on December 11th and 12th, 2000, as the spacecraft neared the planet during a flyby on its way to Saturn (source: NASA/JPL/Space Science Institute [16]).

succinct paper on *Nature* in 1956 [11], ascribed the formation of the Bénard's cells to the variation of the surface tension of the bounding free surface with temperature, i.e. the Marangoni effect. By repeating the linear stability analysis of Rayleigh, Pearson [12] demonstrated that «cellular convective motion of the type observed by H. Bénard (...) can also be induced by surface tension forces». In the latter case, he also determined a critical value of the relevant control parameter, the Marangoni number.

In modern language, the convective motion induced by the variations of the surface tension with temperature is referred to as Bénard-Marangoni (BM) convection; conversely, the convective motion driven by buoyancy forces in a layer of fluid subjected to a temperature gradient parallel to the gravity is called Rayleigh-Bénard (RB) convection. BM convection occurs only in a very thin layer of fluid with an upper free surface (provided that the Marangoni number exceeds a critical value). On the other side, RB convection is observed in a broad range of flow motions in which buoyancy effects have some relevance with respect to viscous/inertia forces.

The ubiquitous character of the RB convection explains the growing and growing interest that such a phenomenon has been arousing throughout the last century.

Not only has it prompted the development of stability theories, as above discussed (for further details the reader is referred to the classical textbooks [17, 18]), it has also become a paradigm for the explanation and understanding of different mechanisms inherent to motions in atmosphere and oceans [19, 20], geophysical and astrophysical flows [21–27] and technological processes [28, 29]. In most of these contexts, the convective motion is also affected by a background rotation, which significantly changes the underlying physics. Thermal convection in presence of rotation is often referred to as rotating RB convection.

The origin of the so-called "trade winds" as a result of the combined action of the buoyancy-driven convection in the atmosphere and Earth's rotation was conjectured by Hadley a long time before the Benard's experiments [19]. Thermal convection is also responsible of the global thermohaline circulation (figure 1.1b), which interacts in a non-linear way with wind-driven oceanic currents (affected by rotation) [20]. Such phenomena have considerable effects on our planet's climate.

On the geophysical side, plate tectonics is associated with Earth's mantle convection [21]. In fact, although predominantly solid, Earth's mantle behaves as a viscous fluid on geologic time scales and RB convection can apply to model its movements. Within such a convective motion, the rising of deep mantle plumes (with a diameter approximately equal to 150 km and a speed of 2m/year) results in superficial hot spots, which, after the plate motion, cause the formation of volcanic island chains and aseismic ridges [22]. In the last decade of the past century, rotating convection models were also used to account for the generation of the Earth's magnetic field [23]. Geodynamo theories thrived on the discovery that the Earth's core is fluid [30]. In an interesting investigation [24], Glatzmaier and Roberts simulated the magneto-hydrodynamic equations in a rapidly rotating spherical fluid shell with a solid conducting inner core and they observed, near the end of their simulation, a magnetic field reversal. This result is considered as an evidence that the rotating magneto-convection models can effectively describe also the magnetic pole inversions, occurred several times in the Earth's history and documented by magnetofossils.

On the astrophysical side, solar granulation, which remarkably resembles the reticulated structures first observed by Bernard in 1900 (figure 1.1c), can be attributed to thermal convection in the Sun's outer layer [25]. On the other hand, solar deeper convection is likely to be more influenced by rotation and to affect Sun's magnetic activity [26]. The emergence of zonal flows in the major planets of the solar system (like Jupiter, see figure 1.1d) exhibits striking similarities with the patterns of the buoyancy-driven flows in rapidly rotating spherical fluid shells [27].

In technological applications, examples of thermal convection influenced by rotation include the cooling process of the turbine blades in turbomachines [28] and the efficient separation of carbon dioxide ($CO_2$) from methane or nitrogen gas [29]. The latter process consists essentially in centrifuging the gas mixture at high pressure: the centrifugal forces cause the $CO_2$ to be concentrated at the walls of the centrifuge and, due to radial compression, the gas is then condensed. Being an exothermic process, condensation is accompanied by heat release and thus a radial temperature gradient adds to the rotation.

In the light of the above-presented variety of applications, a large number of experimental, numerical and theoretical works have been carried out in last years aiming at analyzing and understanding in depth different and relevant aspects of the RB convection. As pointed out by Xia in a recent review article [31], the major directions in the studies of convective thermal turbulence are: turbulent heat transfer, boundary layer dynamics, coherent structures and flow dynamics and small-scale turbulence. The present investigation focuses mainly on the third direction, i.e. investigation of the large scale structures in the flow fields of rotating and non-rotating RB convection.

In the remainder of the present chapter, an introduction to the basic theories and the current status of knowledge on RB convection is given. Such an overview is solely intended to let the reader to familiarize with the concepts invoked in other parts of the dissertation; therefore, it does not cover all the relevant results achieved in the research field of thermal convection. For a more comprehensive introduction to the major issues of RB convection, the reader is referred to the recent reviews by Ahlers *et al.* [32], Lohse and Xia [33] and Xia [31]. A thorough review of rotating RB convection is given by Stevens *et al.* [34].

## 1.2. Governing equations

### 1.2.1. Oberbeck-Boussinesq approximation and rotation effects

RB convection is usually studied by employing the so-called Oberbeck-Boussinesq (OB) approximation [35–37], which holds in most situations where large temperature gradients do not occur. Within the OB simplification, density can be treated as a constant in all the terms in the governing equations of motion, with the exception of the term related to the external forces, i.e. buoyancy. To account for variations of the buoyancy forces, a linear relationship between the density $\rho$ and the temperature $T$ is assumed:

$$\rho(T) = \rho_0[1 - \beta(T - T_0)] \tag{1.1}$$

where $\beta$ is the thermal expansion coefficient, $T_0$ is a reference temperature and $\rho_0$ the corresponding value of density. In addition, all the material properties of the fluid, such as $\beta$, the kinematic viscosity $\nu$ and the thermal diffusivity $\alpha$, are supposed to be constant. With the above assumptions, the governing equations of the motion for a Newtonian fluid are the Oberbeck-Boussinesq equations [37]:

$$\nabla \cdot \mathbf{U} = 0 \tag{1.2}$$

$$\partial_t \mathbf{U} + \mathbf{U} \cdot \nabla \mathbf{U} = -\nabla P + \nu \nabla^2 \mathbf{U} - \beta \Theta \mathbf{g} \tag{1.3}$$

$$\partial_t \Theta + \mathbf{U} \cdot \nabla \Theta = \alpha \nabla^2 \Theta \tag{1.4}$$

where $\mathbf{U}$ is the velocity vector, $P = P_s/\rho_0 - \psi$ is the reduced pressure with $P_s/\rho_0$ being the kinematic pressure and $\psi$ being a potential for the gravitational field ($\nabla \psi = \mathbf{g}$), $\Theta = T - T_0$ and $\partial_t$ denotes the temporal partial derivative.

Equations (1.2-1.4) are valid in any inertial reference frame and still hold when the fluid is affected by a background rotation. In the latter case, rotation enters the governing equations through the boundary conditions, as the boundaries of the fluid domain are rotating with respect to the inertial reference frame. On the other side, the effects of rotation can be easier understood by analyzing the fluid motion directly in the rotating reference frame. This approach requires a transformation of equations (1.2-1.4) from the inertial reference frame to the rotating one. For an incompressible flow, only the momentum equation (1.3) is modified with the introduction of two kinds of apparent forces: the Coriolis forces and the centrifugal forces. In particular, applying the relation:

$$(\partial_t)_{\text{inertial}} = (\partial_t)_{\text{rot}} + \boldsymbol{\Omega} \times \tag{1.5}$$

to the radius vector $\mathbf{R}^1$ yields $\mathbf{U} = \mathbf{U}_{\text{rot}} + \boldsymbol{\Omega} \times \mathbf{R}$ and, applying again equation (1.5) to the velocity vector $\mathbf{U}$, it follows that:

$$\partial_t \mathbf{U} = \partial_t \mathbf{U}_{\text{rot}} + 2\boldsymbol{\Omega} \times \mathbf{U}_{\text{rot}} + \boldsymbol{\Omega} \times (\boldsymbol{\Omega} \times \mathbf{R}). \tag{1.6}$$

In the above equations, $\mathbf{U}$ and $\mathbf{U}_{\text{rot}}$ denotes the velocity in the inertial and rotating reference frames, respectively, whereas $\boldsymbol{\Omega}$ is the angular velocity. In equation (1.6) $2\boldsymbol{\Omega} \times \mathbf{U}_{\text{rot}}$ is the Coriolis acceleration, while $\boldsymbol{\Omega} \times (\boldsymbol{\Omega} \times \mathbf{R})$ is the centrifugal acceleration. It is now worth remarking that within the OB approximation it is not generally acceptable to ignore variations in density due to variations in temperature for the term related to the centrifugal forces, while such variations are negligible for the Coriolis term. By applying the linear relationship (1.1) to account for density variations in the centrifugal force term, equation (1.6) turns into:

$$\rho\partial_t \mathbf{U} \approx \rho_0 \partial_t \mathbf{U}_{\text{rot}} + 2\rho_0 \boldsymbol{\Omega} \times \mathbf{U}_{\text{rot}} + \rho_0 \boldsymbol{\Omega} \times (\boldsymbol{\Omega} \times \mathbf{R}) - \rho_0 \beta\Theta\, \boldsymbol{\Omega} \times (\boldsymbol{\Omega} \times \mathbf{R}). \tag{1.7}$$

The momentum equation in the rotating reference frame can be obtained by replacing the first term in equation (1.3) by equation (1.7) and assuming $\mathbf{U} \equiv \mathbf{U}_{\text{rot}}$ in the remaining terms. Finally, the equation of motions in the rotating reference frame are:

$$\nabla \cdot \mathbf{U} = 0 \tag{1.8}$$

$$\partial_t \mathbf{U} + \mathbf{U} \cdot \nabla \mathbf{U} = -\nabla P + \nu\nabla^2\mathbf{U} - \beta\Theta\mathbf{g} +$$

$$-2\boldsymbol{\Omega} \times \mathbf{U} + \beta\Theta\, \boldsymbol{\Omega} \times (\boldsymbol{\Omega} \times \mathbf{R}) \tag{1.9}$$

$$\partial_t \Theta + \mathbf{U} \cdot \nabla\Theta = \alpha\nabla^2\Theta \tag{1.10}$$

where the term $\boldsymbol{\Omega} \times (\boldsymbol{\Omega} \times \mathbf{R})$ has been absorbed in the reduced pressure term by

---

[1] The radius vector $\mathbf{R}$ is defined as the vector that identifies the position of a point in the fluid domain with respect to a point arbitrarily located on the rotation axis.

using the vectorial identity $2\boldsymbol{\Omega} \times (\boldsymbol{\Omega} \times \mathbf{R}) = \nabla(|\boldsymbol{\Omega} \times \mathbf{R}|^2)$ and the subscript 'rot' has been avoided for clearness. It should be noted that the above equations stands in the case of constant angular velocity, otherwise there occur additional fictitious forces related to the angular acceleration $\partial_t \boldsymbol{\Omega}$, known as Euler forces.

The closure of the problem governed by equations (1.8-1.10) requires the definition of a fluid domain and appropriate boundary conditions. In canonical RB convection, the fluid domain is delimited by two flat surfaces that are horizontal (i.e., normal to $\mathbf{g}$), separated by a distance $L$ and maintained at constant temperature, with the lower one at a temperature $T_b$ greater than that of the upper one $T_t$ (i.e., $\Delta = T_b - T_t > 0$). No-slip condition is applied to both surfaces ($\mathbf{U} = 0$). In principle, the fluid domain should not be confined laterally, so as to present only one direction of inhomogeneity. Although stability studies usually rely on such an assumption, in both experimental and numerical practice it is almost impossible to satisfy a similar condition. In fact, also when in numerical simulations an artificial horizontal periodicity is assumed, the lateral size of the computational domain bounds the lowest frequency that can be reproduced in the flow field. For the above reasons, a large number of recent investigations have focused on the analysis of RB convection in presence of a sidewall; this has made it possible to carry out a direct comparison between experimental, numerical and theoretical results.

The most popular geometry for confined RB convection is perhaps the cylindrical one, which has only one additional direction of (statistical) inhomogeneity (the radial one). This also results in only one more characteristic length, i.e. the cylinder diameter $D$. The boundary conditions commonly imposed at the sidewall are the no-slip condition and the adiabatic condition (i.e., the heat flux across the sidewall is assumed to be zero). As shown below (section 1.6), the latter condition is never completely achieved in a laboratory experiment and, thus, appropriate considerations or corrections are needed when measurements are compared with numerical and theoretical results.

### 1.2.2. Dimensionless parameters

$\mathbf{N}$ON-DIMENSIONALIZATION of equations (1.8-1.10) and of the corresponding boundary conditions leads to the introduction of the main control parameters of RB convection. By using the free-fall scaling, which adopts $L$ as length scale, the so-called free-fall velocity $u_0 = \sqrt{\beta \Delta g L}$ as velocity scale and $\Delta$ as temperature scale, the non-dimensionalized equations of motion are:

$$\nabla \cdot \mathbf{u} = 0 \tag{1.11}$$

$$\partial_t \mathbf{u} + \mathbf{u} \cdot \nabla \mathbf{u} = -\nabla p + \sqrt{\frac{Pr}{Ra}} \nabla^2 \mathbf{u} - \theta \hat{\mathbf{g}}$$

$$- \frac{1}{Ro} \hat{\boldsymbol{\Omega}} \times \mathbf{u} + \frac{2Fr}{\Gamma} \theta \hat{\boldsymbol{\Omega}} \times (\hat{\boldsymbol{\Omega}} \times \mathbf{r}) \tag{1.12}$$

$$\partial_t \theta + \mathbf{u} \cdot \nabla \theta = \sqrt{\frac{1}{PrRa}} \nabla^2 \theta. \tag{1.13}$$

Here, $\hat{\mathbf{g}}$ and $\widehat{\mathbf{\Omega}}$ are the unit vectors parallel to the directions of the gravity and the angular velocity, respectively, and five dimensionless number can be defined, namely the Rayleigh number $Ra$, the Prandtl number $Pr$, the Rossby number $Ro$, the Froude number $Fr$ and the aspect ratio $\Gamma$:

$$Ra = \frac{g\beta\Delta L^3}{\alpha\nu} \tag{1.14}$$

$$Pr = \frac{\nu}{\alpha} \tag{1.15}$$

$$Ro = \frac{\sqrt{\beta\Delta g/L}}{2\Omega} \tag{1.16}$$

$$Fr = \frac{\Omega^2 D}{2g} \tag{1.17}$$

$$\Gamma = \frac{D}{L}. \tag{1.18}$$

In the above equations the non-dimensional quantities are denoted with the lower-case variants of the symbols used for the corresponding dimensional quantities in equations (1.8-1.10).

In canonical rotating RB convection, it is supposed that $\widehat{\mathbf{\Omega}} = \pm\hat{\mathbf{g}}$, while the effects of centrifugal forces are typically neglected. From equation (1.12) this is clearly possible only when $Fr \ll 1$. At a closer look, the adopted scaling suggests that centrifugal effects are important for very slender cylinders, i.e. $\Gamma \ll 1$, which seems to be contrary to the physical expectations. Such a contradiction comes from the fact that the non-dimensional centrifugal acceleration is, in general, not unit order. This is evident when the term $\widehat{\mathbf{\Omega}} \times (\widehat{\mathbf{\Omega}} \times \mathbf{r})$ is written in the form $-r\hat{\mathbf{r}}$, with $r$ being the (non-dimensional) distance from the rotation axis and $\hat{\mathbf{r}}$ the unit vector in the radial direction. Since the maximum value of $r$ is $0.5\,\Gamma$, the term $2\Gamma^{-1}\widehat{\mathbf{\Omega}} \times (\widehat{\mathbf{\Omega}} \times \mathbf{r})$ is at most equal to unity and, thus, the Froude number defined in equation (1.17) indeed determines the order of magnitude of the term related to the centrifugal forces.

The Rossby number is not the only possible choice to nondimensionalize the angular velocity $\Omega$. Among the most common alternatives employed in literature are the Taylor number $Ta$ and the Ekman number $Ek$:

$$Ta = \left(\frac{2\Omega L^2}{\nu}\right)^2 \tag{1.19}$$

$$Ek = \frac{\nu}{\Omega L^2}. \tag{1.20}$$

While the Rossby number compares the Coriolis forces to the buoyancy ones, the Taylor and the Ekman numbers compare the Coriolis forces to the viscous ones. Note also that $Ta = (2/Ek)^2$.

In addition to the above-defined control parameters, further non-dimensional parameters can be introduced to characterize the global response of the RB system. A first key parameter is represented by the area- and time-averaged total heat flux $\dot{q}$ across any horizontal section of the fluid domain. With an adiabatic sidewall, such a quantity is independent of the vertical coordinate. In non-dimensional terms, $\dot{q}$ can be expressed as the Nusselt number:

$$Nu = \frac{\dot{q}L}{\kappa\Delta} \qquad (1.21)$$

with $\kappa$ being the thermal conductivity of the fluid. From equation (1.13), it follows that:

$$Nu = \sqrt{PrRa}\,\langle \overline{u\theta}\rangle_{y,z} - \partial_x\langle\overline{\theta}\rangle_{y,z} \qquad (1.22)$$

where $\langle\cdot\rangle_{y,z}$ denotes the average over a horizontal plane and over time, $u$ is the vertical velocity component, $\partial_x$ indicates the spatial derivative in the vertical direction and both velocity and temperature are non-dimensional. Focusing on equation (1.21), it is interesting to note that the quantity $\kappa\Delta/L$ is the pure conductive heat flux obtained in absence of the fluid motion; therefore, $Nu$ is equal to the unity in the conductive case, and greater when thermal convection occurs. On the other side, the quantity $\dot{q}L/\Delta$ may be interpreted as the effective thermal diffusivity $\kappa_{\text{eff}}$ of the convective layer. This leads to the conclusion that in the pure conductive case $\kappa_{\text{eff}} = \kappa$, otherwise $\kappa_{\text{eff}} > \kappa$.

The second key parameter of the RB convection is a Reynolds number defined as:

$$Re = \frac{UL}{\nu} \qquad (1.23)$$

where $U$ is a characteristic velocity scale of the turbulent motion. Different definitions for this characteristic velocity scale have been proposed, as reported in the Ahlers *et al.*'s review [32]. Basically, it is possible to distinguish between two different types of Reynolds numbers: those based on the large scale circulation (LSC) and those based on the plume motion. Some discussion about these dynamical structures of the RB convection is carried out in section 1.4.

## 1.3. Thermal wind balance and Ekman pumping

IN this section, it is shown, by virtue of appropriate simplification of equations (1.11-1.13), that rotation can significantly affect the dynamics of the thermal turbulence, introducing important effects both in the bulk and in proximity of the boundary layers that form on the walls of the cell. Classical textbooks dealing with the effects of rotation are [38, 39].

Let us suppose that $Ro$ is small enough to make convective accelerations $(\mathbf{u}\cdot\nabla\mathbf{u})$ negligible with respect to the Coriolis ones and $Fr \ll 1$. Sufficiently far from the walls, the effects of the fluid viscosity can be neglected and, if the flow is steady, the momentum equation (1.12) simplifies to:

$$\frac{1}{Ro}\mathbf{u} \times \widehat{\mathbf{\Omega}} = \nabla p + \theta \hat{\mathbf{g}}. \tag{1.24}$$

Supposing that the rotation axis is vertical ($\widehat{\mathbf{\Omega}} = -\hat{\mathbf{g}}$) and coincident with the $x$-direction of an arbitrarily chosen Cartesian reference frame $Oxyz$, the horizontal components of equation (1.24) read as:

$$w = Ro\,\partial_y p \tag{1.25}$$
$$v = -Ro\,\partial_z p \tag{1.26}$$

which constitutes the so-called geostrophic balance, a balance between the pressure gradient and the Coriolis force that makes streamlines coincident with isobars (from the above equations pressure is in fact a streamfunction for the flow in the horizontal plane). Now, applying the rotor operator to equation (1.24) yields (after some algebra):

$$\partial_x \mathbf{u} = Ro\,\nabla\theta \times \hat{\mathbf{g}} \tag{1.27}$$

which represents the so-called thermal wind balance. The projections of equation (1.27):

$$\partial_x u = 0 \tag{1.28}$$
$$\partial_x v = -Ro\,\partial_z \theta \tag{1.29}$$
$$\partial_x w = Ro\,\partial_y \theta \tag{1.30}$$

show that, in such a regime, the vertical gradients of the vertical velocity are zero (although in general $u$ can be non-zero) and the vertical gradients of the horizontal velocity are in fact related to horizontal gradients of temperature. When only vertical gradients of temperature are present (i.e., $\nabla\theta$ is parallel to $\hat{\mathbf{g}}$) no vertical gradients of the velocity vector are possible. This happens as well when $Ro \ll 1$, as clearly shown by equation (1.27). Such a result is known as Taylor-Proudman theorem and states that a steady, slow and inviscid flow in a rotating fluid is indeed two-dimensional in the plane perpendicular to the rotation axis. The Taylor-Proudman theorem explains, on one side, the formation of columnar structures in strongly rotating thermal convection (see section 1.4), on the other, the drop of turbulent heat transfer with at very small $Ro$, which was observed since the first experiments of Rossby in 1969 [40] (see section 1.5).

If the geostrophic balance can apply to the bulk motion, where the effects of viscosity are in fact negligible, the boundary layers on the top and the bottom plates can be effectively modelled by the Ekman boundary layer theory. Ekman layers are boundary layers induced by small differences in relative rotation between the fluid

and the wall, occurring in a quasi-steady flow at $Ro \ll 1$. These boundary layers have
a thickness independent of the radial and azimuthal coordinates and determined only
by the two relevant quantities of the flow phenomenon, i.e. the kinematic viscosity
and the angular velocity. Leaving aside the mathematical treatment of the theory, it
can be demonstrated that the Ekman boundary layer is characterized by a vertical
velocity profile given by:

$$u = \frac{1}{2}\omega_b\delta_E \left\{ 1 - \exp\left(-\frac{x}{\delta_E}\right) \left[ \sin\left(\frac{x}{\delta_E}\right) + \cos\left(\frac{x}{\delta_E}\right) \right] \right\} \qquad (1.31)$$

where $\omega_b$ is the $x$-component of the vorticity at the edge of the boundary layer (i.e.,
in the bulk motion) and $\delta_E = \sqrt{\nu/\Omega}$ is the thickness of the Ekman boundary layer.
Note that $\delta_E = L\sqrt{Ek}$. Equation (1.31) shows the strict dependence of the motion
in the Ekman boundary layer on the bulk flow. In particular, positive vorticity $\omega_b$
at the edge of the boundary layer induces an ascending flow in the Ekman layer
(mechanism known as Ekman pumping), while negative $\omega_b$ causes descending flow
(mechanism known as Ekman suction). As below commented, Ekman pumping is
responsible of the increase of $Nu$ at moderate values of $Ro$ (see section 1.5).

## 1.4. Flow dynamics and coherent structures

ONE of the first and undoubtedly most detailed descriptions of the coherent
structures in the RB convection is due to Zocchi *et al.* [41]. They used the liquid
crystals visualization technique to study non-rotating thermal convection in a cubic
box filled with water at $Ra = 1.2 \times 10^9$ and $Pr = 5.6$, insulated from the outside
with styrofoam. In the epilogue of their article, referring to the complex organization
of the turbulent flow investigated, they wrote: «It is a small world of its own, with
clouds and rain, wind and storms». This is also one of the most suggestive and
concise definition of confined thermal turbulence.

Zocchi *et al.* [41] identified essentially five kinds of coherent structures, namely
the large scale circulation (LSC), the plumes, the thermals, the waves and the
spiraling swirls. Figure 1.2 is a schematic representation of such structures, adapted
from figure 3 of the Zocchi *et al.*'s article. The LSC (also known as wind of the
turbulent convection) is a domain-filling circulatory motion that exhibits a "flywheel"
structure confined mainly in a near-vertical plane. In the case of a rectangular
domain, this plane coincides with one of the two diagonal planes, as shown in figure
1.2. Plumes are identified both as columns of hot (cold) fluid that rise up (fall down)
from the thermal boundary layers due to buoyancy forces and as sheet-like structures
that move horizontally near the boundary layers, driven by the LSC. Vertical plumes
can be either attached to the boundary layers where they form or be detached and, in
the latter case, they are called thermals (or often "puffs"). The clustering of plumes
on the sides of the LSC is often referred to as "jet" [42]. When plumes or thermals,
in the form of jets, hit the boundary layer on one plate, they produce waves that
are sustained by the wind of the convection. During their propagation, waves can

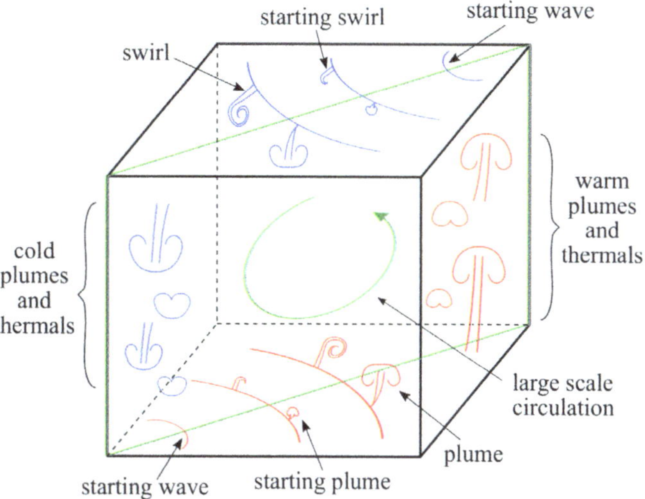


Figure 1.2: Sketch of the coherent vortex structures in thermal convection. Adapted from [41].

result in new plumes, if buoyancy effects dominate over inertia, or in spiraling swirls, in the opposite case. Zocchi *et al.* [41] observed that swirls are very reminiscent of the vortices in the Kelvin-Helmholtz instability, typical of free-shear flows. In the outlined picture, the release of plumes is triggered by waves produced by the arrival of plumes at the boundary layer; these plumes, in turn, will rise up or fall down to the opposite boundary layer and excite new waves, thus sustaining the life cycle of the turbulent convection.

While in this life cycle the interplay between the LSC and the plumes is straight-forward, what is less evident is the mechanism leading to the onset of the LSC. In a combined shadowgraph and particle image velocimetry (PIV) investigation, Xi *et al.* [43] found that the initial horizontal motion required by the LSC is subsequent to the entrainment caused by the plume's vertical motion. Such an entrainment causes the formation of vortices and the gradual merging of plumes and vortices results in a coherent circulatory motion mainly consisting of the plumes themselves and spanning the whole convection box.

The dynamics of the flow structures in a cylindrical geometry is even more complex than that of a rectangular box. While a quasi-planar LSC is still observed in this case, the rotational invariance of the governing equations implies the ax-isymmetry of the statistically-average flow field, which means that there exists no preferential azimuthal orientation of the LSC plane. Based on this, one expects a continual re-orientation of the LSC plane, which, in fact, has been observed in several experimental investigations [44–49]. Such a rotation is a spontaneous diffusive meandering characterized by a linear increase of the azimuthal angle with the time. Re-orientations add to other interesting oscillatory modes, namely a torsional mode [50, 51], a sloshing mode [52, 53] and cessations and reversals (i.e., abrupt interruptions and changes in orientations) of the circulatory direction [47, 49, 54]. Due to the torsional mode, LSC undergoes azimuthal oscillations that are out of phase between the upper and lower halves of the cell. On the other side, the sloshing mode consists of an in-phase horizontal displacement of the entire LSC and is associated with horizontal fluctuations of both velocity and temperature in the bulk. Before the discovery of such a mode, these oscillations had been associated to the periodic and alternate emissions of plumes from the opposite boundary layers [55]. However, the work by Xi *et al.* [52] elucidated conclusively that thermal plumes are emitted neither periodically nor alternately, but randomly and continuously, from the top and bottom plates. Cessations and reversals of the thermal convection have been often associated with Earth's magnetic pole inversion [24] and also reversals of the wind in Earth's atmosphere [56]; hence, the great interest of the scientific community in determining the features of their statistical occurrence. Studies in this direction have shown that, after a cessation, any azimuthal orientation has the same probability to occur and, thus, reversal is only a special case of cessation. Moreover, since cessation events are Poisson distributed in time, successive cessations are statistically uncorrelated [54].

Most of investigations mentioned above are related to thermal convection inside cylinders with aspect ratio equal to 1. In slender cylinders ($\Gamma < 1$), the LSC has been observed to switch between different states. In their direct numerical simulations on thermal turbulence in a cylindrical sample with $\Gamma = 1/2$, Verzicco and Camussi [57] found out the existence of a flow mode consisting of two vertically-stacked near-circular counterrotating rolls, which were later observed experimentally by Xi and Xia [58]. The latter study revealed random temporal successions of one-roll and double-roll states, with a prevalent occurrence of the two-roll state as $\Gamma$ is decreased. In [59] a simple model for the prediction of this bimodality was also given.

As one can expect, the addition of rotation significantly alters the flow dynamics of the RB convection. As discussed in more detail in section 1.5, also the global heat transfer is affected to large extent by the effects of rotation and, when the Nusselt number in presence of rotation is compared to that in absence of it, three different regimes can be detected, corresponding to weak, moderate and strong rotation, respectively.

In the regime of weak rotation, the LSC is still present, but its dynamical properties are considerably affected by rotation [60–63], even though the global heat transfer is nearly unchanged with respect to the non-rotating case. From a dynamical viewpoint, the two perhaps most interesting features that have been observed are the precession of the LSC in the direction opposite to the background rotation [60, 61] and the increase in the frequency of cessations [62]. Although some models have been developed to predict these phenomena, a clear understanding of the physical mechanisms from which they origin is still missing. Further relevant characteristics of this regime are the increase in the temperature amplitude of the LSC and the decrease of the temperature gradient along the sidewall [63].

The regimes of moderate and strong rotation are characterized by the breakdown of the LSC and the emergence of vertically-aligned vortices, often denoted as Taylor-Proudman columns. Figure 1.3 illustrates how the morphology of the coherent vortex

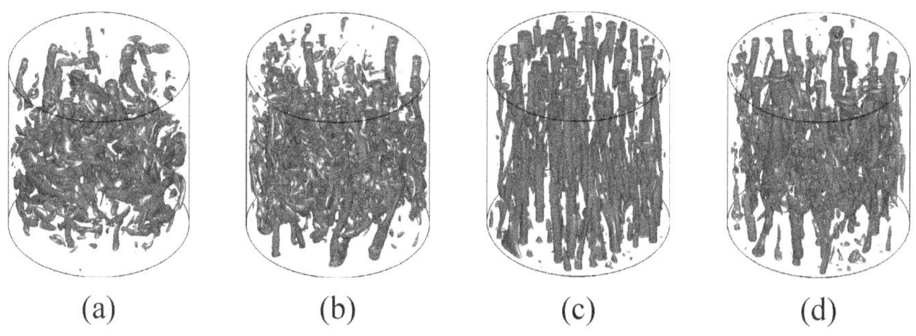

Figure 1.3: Evolution of the coherent vortex structures with increasing Taylor number. (a) $Ta = 3.0 \times 10^8$; (b) $Ta = 1.2 \times 10^9$; (c) $Ta = 4.8 \times 10^9$; (d) $Ta = 1.9 \times 10^{10}$. Adapted from [64].

structures, identified by the three-dimensional $Q$-criterion [65, 66], varies within these regimes as rotation (expressed in terms of the Taylor number) increases. At $Ta = 3.0 \times 10^8$ the flow consists of an intricate network of elongated vortical tubes, which gradually turn into vertically stretched vortices for larger $Ta$. Several works [64, 67, 68] have shown that the number of the columnar vortices near the plates (or better, at the edge of the thermal boundary layers) greatly increases passing from the first regime to the second one. Moreover, vortex concentration is reduced in the region adjacent to the sidewall, while a uniform and regular pattern is found in the center. It should be noted, however, that the columnar vortices do not extend from the bottom to the top: indeed, they lose gradually vorticity entering the bulk and experience a spin-down approaching the opposite plate [64].

The behavior of rotating plumes shows features significantly different from that of the non-rotating case. If on one side the entrainment of plumes is suppressed by rotation (according to the Taylor-Proudman theorem), on the other side the mutual interaction of plumes leads to a mixing of their properties with the bulk fluid without any change in the volume of the plumes themselves [69, 70]. The mixing tends to reduce the plume buoyancy anomaly, leading to a decrease in the total heat transported by a plume as it migrates across the layer.

Another relevant feature of the regimes of moderate and strong rotation is the presence of a strong temperature gradient at the sidewall. The latter has been associated to a recirculation in the Stewartson boundary layer on the sidewall with upward (downward) transport of hot (cold) fluid close to the sidewall in the bottom (top) part of the cell [71]. Such a secondary flow is generated by the Ekman pumping, which is also responsible of the strong increase in the vertical velocity fluctuations at the edge of the thermal boundary layer [67, 68]. For very strong rotation, the vertically aligned vortices tend to merge and this also leads to an intense destabilizing temperature gradient in the bulk [69, 70, 72–75].

# 1.5. Turbulent heat transfer

ONE of the main issues of the research on RB convection, to which a great amount of work has been devoted, is to determine how the turbulent heat transfer depends on the control parameters, i.e. the identifcation of the functional relationship $Nu(Ra, Pr, Ro, \Gamma)$.

For non-rotating RB convection (i.e., in the limit of $Ro \to \infty$), among the several theoretical models that have been proposed over the years, the Grossman and Lohse (GL) theory [76–79] seems to be the most successful one. The basic idea of this theory is to split the volume- and time-averaged kinetic and thermal dissipations $\varepsilon_u$ and $\varepsilon_\theta$ in two contributions, one related to the bulk motion (i.e., the large scale circulation) and the other related to the boundary layer (i.e., the plumes). The contributions of the first type are modeled according to the Kolmogorov's energy-cascade picture, thus assuming isotropic and homogeneous turbulence in the bulk, while the boundary layer contributions are modeled by the Prandtl-Blausius *laminar* boundary layer theory for steady flows over an infinite flat plate. Four different regimes are then detected (depending on which of the two contributions dominates the terms $\varepsilon_u$ and/or $\varepsilon_\theta$) and, for each of them, different exponents in the $Ra$ and $Pr$ scalings of $Nu$ and $Re$ are found.

The most arguable point of the GL theory is its capability to predict correctly the turbulent heat transfer in the so-called "ultimate", or "asymptotic", regime, reached at high $Ra$ and low to moderate $Pr$. In the ultimate regime, which was first postulated by Kraichnan in 1962 [80] and Spiegel in 1971 [81], the heat flux $\dot{q}$ does not depend anymore on the molecular diffusivities of the fluid and this should be possible as a result of the transition of the boundary layers to a fully turbulent state. In the GL theory the ultimate regime is predicted as one of the several states, where the $Ra$ and $Pr$ dependencies for $Nu$ and $Re$ are given by: $Nu \sim Ra^{1/2}Pr^{1/2}$ and $Re \sim Ra^{1/2}Pr^{-1/2}$. Kraichnan's model predicts a different exponent for the $Pr$ dependencies in the range $0.15 < Pr \leq 1$ and logarithmic corrections for the $Ra$ dependencies. So far, experiments approaching the ultimate regime have been performed with helium gas operated near its critical point at very low temperatures [82–84] or with pressurized gases close to ambient temperatures [85]. However, the almost controversial results of such investigations leave open the question about the existence of the ultimate regime.

As concerns the $Pr$ dependence of $Nu$ at small or moderate $Ra$, several measurements with mercury [40, 44, 86], liquid sodium [87] or helium gas [82–84], but also water, have proved a strong increase for $Pr < 1$ (where the GL theory predicts a scaling with an exponent of 1/8, in good agreement with experiments) and a saturation for greater values of $Pr$. For even larger $Pr$, as those obtained in experiments with organic fluids [88, 89], a gradual decrease of $Nu$ with $Pr$ is found, which is, once more, successfully described by the GL model.

Although the dependencies of $Nu$ and $Re$ on $Ra$ and $Pr$ are effectively caught by the GL theory, the latter neglects the influence of $\Gamma$ on the response parameters. Indeed, the main ingredient of the GL model is the existence of a large-scale circulatory motion (the LSC), well distinguished from the boundary layers. Several

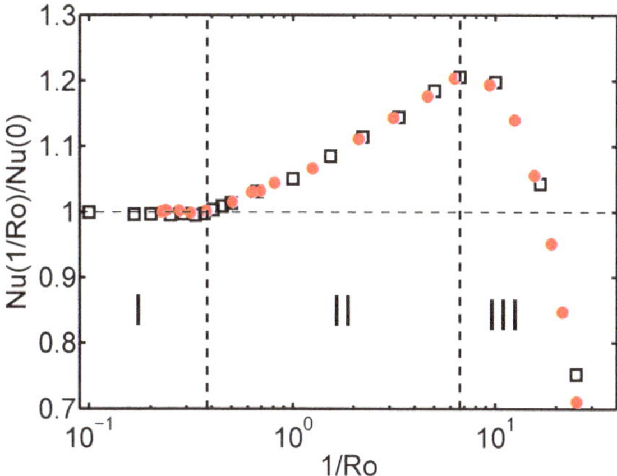

Figure 1.4: Normalized heat transfer as a function of $1/Ro$ on logarithmic scale. The red dots and black squares correspond to experimental and numerical data, respectively, in the case $Ra = 2.73 \times 10^8$ and $Pr = 6.26$ for $\Gamma = 1$. The vertical dashed lines demarcate the three different regimes of rotating RB convection. The picture is taken from Stevens *et al.*'s review paper [34].

investigations [90, 91] have shown that very different flow structures are observed in cells of different $\Gamma$ and that the domain-filling LSC is replaced by multi-roll structures as $\Gamma$ roughly exceeds 4 [92]. Nonetheless, $Nu$ has been found to exhibit a very weak dependence on $\Gamma$ in both cylindrical and rectangular samples [93–95]. This ultimately suggests an insensitivity of the heat flux to the actual configuration of the coherent vortex structures present in the flow field.

The addition of rotation deeply modifies the above-discussed behavior of the global heat transfer. Specifically, as aforementioned, rotation introduces three different regimes that are reflected in the variation of $Nu$ with $Ro$. These regimes are identified in figure 1.4, where the behavior of the Nusselt number as a function of the inverse Rossby number $1/Ro$ (which is proportional to $\Omega$) is reported.

In the first regime (weak rotation), the turbulent heat transfer is substantially unaffected by the rotation, i.e., $Nu$ keeps approximately the same value as in the non-rotating case. The flow field is still characterized by the existence of a domain-filling LSC, although, as discussed in the previous section, its dynamical properties are considerably influenced by rotation. Conversely, in the second regime (moderate rotation), increasing the rotation rate leads to an enhancement in the global heat flux. This circumstance was already noted by Rossby in his experiments on rotating RB convection [40]. Rossby had measured an increase up to 10%, using water as working fluid. Such an increase seems counterintuitive in the light of the results of the stability analysis of Chandrasekhar [96], which shows that the onset of convection is indeed delayed by rotation. Several works have identified the Ekman pumping as the mechanism responsible for such an increase [40, 97, 98]: associated with the detachment of plumes from the thermal boundary layers, the Ekman pumping determines the formation of the vertically aligned vortices, inside which hot fluid is transferred directly from the bottom plate to the top plate. Finally, the third

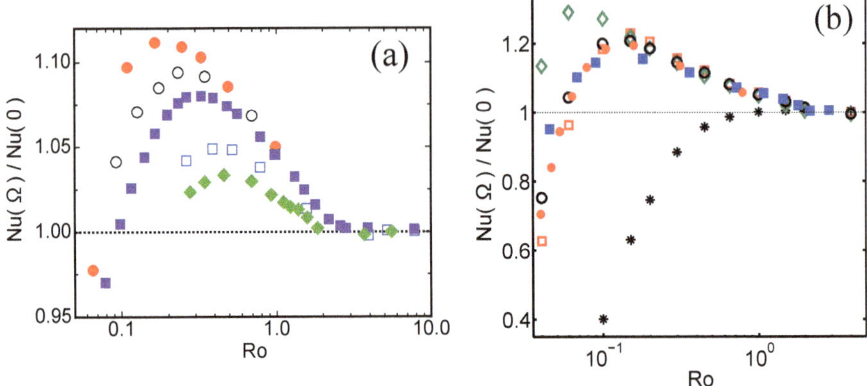

Figure 1.5: Normalized heat transfer as a function of $Ro$ on logarithmic scale. (a) Results at different Rayleigh numbers with fixed Prandtl number, $Pr = 4.38$ (water at 40°C). Red solid circles: $Ra = 5.6 \times 10^8$; black open circles: $Ra = 1.2 \times 10^9$; purple solid squares: $Ra = 2.2 \times 10^9$; blue open squares: $Ra = 8.9 \times 10^9$; green solid diamonds: $Ra = 1.8 \times 10^{10}$. (b) Results at different Rayleigh and Prandtl numbers. Black stars: $Ra = 1 \times 10^8$ and $Pr = 0.7$ (DNS); red solid circles: $Ra = 2.73 \times 10^8$ and $Pr = 6.26$ (experiment); black open circles: $Ra = 2.73 \times 10^8$ and $Pr = 6.26$ (DNS); red open squares: $Ra = 1 \times 10^8$ and $Pr = 6.26$ (DNS); blue solid squares: $Ra = 1 \times 10^9$ and $Pr = 6.26$ (DNS); green open diamonds: $Ra = 1 \times 10^8$ and $Pr = 20$ (DNS). Data taken from [102].

regime is characterized by an abrupt decrease of $Nu$ with $1/Ro$, which is associated to the suppression of the vertical velocity fluctuations as predicted by the geostrophic balance (see section 1.3). Indeed, different turbulent states are observed in the latter regime, also depending on the value of $Pr$ [99, 100].

The above-mentioned behavior of $Nu$ as a function of $Ro$ is not universal. In fact, the values of $1/Ro$ delimiting the different regimes have been found to depend on the other control parameters. Weiss *et al.* [68], based on both experiments and direct numerical simulations, explained the $Nu$ enhancement in the second regime as a finite size effect occurring at a critical value of the inverse Rossby number $1/Ro_c$ inversely proportional to the aspect ratio, i.e., $1/Ro_c \propto 1/\Gamma$. This means that in the case of a horizontally unbounded domain ($\Gamma \to \infty$) the first regime does not exist. Moreover, for stronger rotation, i.e. $1/Ro \gg 1/Ro_c$ the heat transport becomes independent of $\Gamma$ [101]. The latter result has been justified considering that in the rotating regime the heat transfer is essentially related to the vertically aligned vortices, which are local structures rather than domain-filling ones.

Figure 1.5 summarizes the influence of $Ra$ and $Pr$ on $Nu$ in the case of rotating RB convection [61, 102, 103]. In particular, figure 1.5a reports different curves of the ratio of $Nu$ in presence of rotation to $Nu$ in absence of it corresponding to a unique value of $Pr$ and different values of $Ra$ ranging from $5.6 \times 10^8$ and $1.8 \times 10^{10}$. It is possible to see that, as $Ra$ is increased, the second regime tends to cover a smaller range of $Ro$ and also the maximum of the heat transfer enhancement diminishes, while the Rossby number at which this maximum is achieved tends to increase. Figure 1.5b suggests an analogous trend when $Pr$ is decreased. Interestingly, for $Pr = 0.7$ the second regime is not found: this can be associated with the larger thermal diffusivity at low values of $Pr$ which makes the Ekman pumping inefficient (the heat that is carried by the vertical vortices spreads out in the middle of the cell). Indeed, Stevens *et al.* [104] found even for large $Pr$ a reduced effect of the Ekman pumping, associated with the fact that the columnar vortices do not reach the thermal boundary layer, which is, at high $Pr$, much thinner than the kinetic one. Therefore, they concluded that the heat transfer enhancement has a maximum as a function of $Pr$.

## 1.6. Effects of finite conductivity of plates and side-walls

$S$ INCE the current dissertation presents the results of a combined experimental and numerical investigation of RB convection, it is useful to mention here that some discrepancies between measurements and computational simulations arise as a consequence of the physical properties of the sidewall and the finite conductivity of the bottom and the top plate. These topics have been extensively addressed in the case of non-rotating convection [54, 105–111], while smaller attention has been paid in the rotating case.

As explained in section 1.2, in canonical RB convection the top and bottom plates are supposed to be at constant temperature, whereas the sidewall is adiabatic. The constant-temperature condition requires, theoretically, an infinite thermal conductivity of the plates, since each local temperature variation due to the dynamics of the fluid flow has to be counteracted instantly (or at least within a very short time compared to the characteristic times of the convective motion) by the thermal sources. If $\kappa_p$ is the plate conductivity and $t_p$ its thickness, this requirement is equivalent to a large value of the non-dimensional number $X_p = \kappa_p L/(Nu\,\kappa\,t_p)$, where $Nu\,\kappa$ is in fact the effective thermal conductivity $\kappa_{\mathrm{eff}}$ of the convective layer. Note that $R_p = t_p/\kappa_p$ can be interpreted as the thermal resistance of the plate (since the temperature drop across the plate subjected to a steady heat flux $\dot{q}$ is $\Delta_p = R_p\dot{q}$) whereas $R_f = L/\kappa_{\mathrm{eff}} = L/(Nu\,\kappa)$ is the thermal resistance offered by the layer of fluid. Therefore, $X_p \gg 1$ equals to the requirement $R_p \ll R_f$, in such a way that $\Delta_p \approx 0$. When such a condition is not satisfied, the deficiency (excess) of enthalpy caused by the emission of a plume from the bottom (top) boundary layer leaves a cold (warm) spot where the probability of a plume emission is diminished, until this spot is diffused away. This process, of course, affects considerably not only the flow dynamics but also the global heat transfer.

As noted by Verzicco [107], as the Rayleigh number increases, the effect of finite conductivity of the plates becomes the bottleneck of the system, since the increase of $Ra$ results in a heat transfer enhancement, i.e. an increase of $Nu$, which corresponds to a decrease of $R_f$. At very high $Ra$ one can expect that $R_p \sim O(R_f)$ or even that $R_p$ is greater enough than $R_f$ to dominate the thermal resistance of the whole experimental setup. Verzicco [107] performed direct numerical simulations to evaluate the effects of the plate thermal properties on the turbulent heat transfer and determine a correction for $Nu$. On the experimental side, Brown *et al.* [54] carried out experimental measurements for both copper and aluminum plates and were able to determine the Nusselt number $Nu_\infty$ corresponding to infinite thermal conductivity of the plates by extrapolating an empirical formula relating $Nu_\infty$ to the measured Nusselt number $Nu$.

As concerns the influence of the plate diffusivity on the flow dynamics, Hunt *et al.* [106] observed that elongated plumes are the dominating structures of thermal turbulence when $\kappa_p \gg \kappa$, whereas, when $\kappa_p \leq \kappa$, these plumes break into elongated puffs, and for $\kappa_p \ll \kappa$ only small-scale puffs can form near the plate.

If the finite conductivity of the plates plays a fundamental role in the range of large Rayleigh numbers, the influence of the sidewall properties is relevant at small or moderate values of $Ra$. In principle, the adiabatic condition is imposed at the internal side of the lateral wall (i.e., at the surface delimiting the fluid domain), but, in practice, it is possible at most to thermally insulate the external side. Therefore, the lateral wall represents a solid domain of thickness $t_s$ and thermal conductivity $\kappa_s$, thermally coupled with the fluid in the convection cell. The heat transfer between the solid wall and the fluid is negligible only if the ratio $X_s = \kappa_s L/(Nu_l \kappa t_s) = R_f/R_s$ is sufficiently small, with $Nu_l$ being the lateral Nusselt number at a certain height. Indeed, the greater influence of the sidewall on the heat transfer has been observed at small Rayleigh numbers. Several investigations have shown that corrections up to 20%–25% for $Ra = 10^6$ are needed [108–110].

In early works, corrections of $Nu$ for the sidewall effect were assumed negligible or estimated by subtracting the heat transfer for the empty cell. As first noticed by Ahlers [108], this approach is not a good approximation since the wall shares with the fluid, by virtue of the thermal BLs, a large vertical temperature gradient near the top and bottom and a much smaller gradient in the center. This causes a heat current entering the wall at the cell bottom and an equal but opposite current at the cell top; such currents are usually much larger than those for an empty cell. Indeed, Verzicco [110] found that the heat travelling from the hot to the cold plates directly through the sidewall is negligible compared with the heat exchanged at the lateral fluid/wall interface.

It is also worth noting that, although the coupling between the sidewall and the convecting fluid is likely to influence the structure and intensity of the LSC [111], the global heat transfer is determined primarily by processes within the top and bottom boundary layers and thus it is not much affected by the later heat transfer occurring at the fluid/wall interface. In other words, $Nu$ is only determined by the amount of heat that enters the fluid at the bottom plate.

In a more recent study, Stevens *et al.* [112] investigated the influence of the boundary conditions at the external side of the sidewall on the heat transfer. They showed that, by imposing an isothermal condition with the temperature of the external side $T_e$ equal to the mean temperature $T_m$ inside the convection cell, a higher heat transfer is obtained at lower $Ra$, because part of the heat current circumvents the thermal resistance of the fluid by going through the sidewall. However, this effect diminishes at higher $Ra$. Moreover, in agreement with previous experimental results [85], at $T_e > T_m$ they observed a lower heat transfer at the bottom plate and also found that this decrease is not a function of $Ra$. In the same work, it was noticed that different properties of the sidewall can determine different flow organizations and, correspondingly, different heat transport.

## 1.7. Motivation and structure of the thesis

As outlined in the previous sections, a considerable number of experimental, numerical and theoretical studies have been devoted to the investigation of different aspects of RB convection both in absence and in presence of a background rotation. A great focus has been given on the characterization of the system response in terms of global parameters, like the Nusselt number, as a function of the input parameters. On the other side, attention has been paid to the flow field and its relationship with the overall heat transfer. On the experimental side, the dynamical modes of the turbulent convection have been inspected by either the multi-thermal-probe method [46, 51, 93] or velocity field measurements [45, 58, 95].

The first approach consists in measuring the azimuthal temperature profiles over different levels of the convection cell and infer, indirectly from them, information about the flow evolution. Although such a method is suitable for measurements over considerably long times (in some cases, weeks) with the advantage of building well-converged, robust statistics, it could result in misinterpretation of the character of the large-scale flow [101].

Conversely, anemometric techniques, such as laser Doppler anemometry (LDA) and PIV, have the advantage of providing a clear and often unambiguous picture of the velocity field. However, due to inherently three-dimensional nature of the phenomenon, these techniques allow only a limited description of the flow dynamics. For instance, in their investigation Sun *et al.* [95] tilted the convection cell by a small angle to lock the LSC in a specific plane, so as to analyze the properties of the velocity field in different planes with respect to the LSC by planar PIV. Although this approach has proven useful to investigate the release of thermal plumes from the boundary layers on the opposite plates of the cylindrical sample and their interaction with the bulk turbulent flow, the inclination is known to prevent reorientations of the LSC, lower the global heat transfer with respect to the leveled case and also make a selection between different possible flow states [95, 113]. In fact, the evolution of the thermal convection in a leveled cylinder turns out to be significantly more complex due to the statistical azimuthal symmetry of the flow field and its chaotic behavior, as already discussed. A complete characterization of the oscillatory modes of the LSC is therefore possible only through full-field three-dimensional measurements.

To the author's knowledge, in literature on RB convection, the number of ex-perimental works featuring three-dimensional measurements of the velocity field is limited. Some examples are [114–118]. However, in none of these investigations 3D velocity or acceleration measurements have been performed in the whole domain of the turbulent convection. The first goal of the present work is to design an experi-mental apparatus for carrying out *full-field* velocity measurements of RB convection inside a cylinder by means of tomographic particle image velocimetry (T-PIV). Then, thermal convection in both non-rotating and rotating reference frame is studied in operating conditions largely investigated in literature. The employed methodology offers the possibility to investigate the chaotic dynamics of the flow in a quantitative and visual way overcoming the inherent limits of the 2D two-components PIV and other similar techniques. For the first time, to the author's knowledge, the characteristic modes of the turbulent convection are extracted with state-of-art modal analysis techniques (namely, proper orthogonal decomposition) and the statistical behavior of the large-scale structures is inspected using such modes. The experimental investigation is carried out in conjunction with direct numerical simulations; this allows, on one side, to validate the physical models and computational approaches used to simulate the phenomenon in a numerical environment, on the other side, to point out unavoidable non-idealities of the experimental setup that makes the phenomenon to differ from the canonical problem addressed not only numerically, but also theoretically.

In the following, we first present a literature overview of the current state of the T-PIV technique in chapter 2, focusing on the main issues related to the application of such a technique to the analysis of RB convection. The advantages of some cutting-edge approaches, such as the motion-enhanced reconstruction methods for time-resolved T-PIV and the shake-the-box technique for 4D particle tracking velocimetry, are also presented and discussed. The design of the experimental apparatus, the details of the image processing techniques employed and the numerical methods and procedures are described in chapter 3. Chapter 4 addresses the issue of managing the effects of the optical distortions caused by the cylinder curved sidewall on the tomographic reconstruction. We introduce an innovative camera calibration model, which ensures high accuracy in tomographic reconstruction and thus velocity estimation when the cylinder interior is imaged through the later wall. Chapter 5 reports the experimental results of non-rotating RB convection in a cylinder with aspect ratio equal to 1/2 at $Ra = 1.8 \times 10^8$ and $Pr = 7.6$ in non-rotating frame; a comparison with the numerical results of the DNS and other results available in literature is also carried out. In chapter 6, numerical simulations and experimental measurements of rotating RB convection are comparatively presented and plausible reasons of the observed experiment/DNS mismatch are discussed. Finally, conclusions and future perspectives are drawn in the final chapter.

**2**

# Tomographic particle image velocimetry

As a non-intrusive, whole-field technique for measuring velocities in small to considerably large domains of complex flows, particle image velocimetry (PIV) has evolved, over the last four decades, to perhaps the principal tool to experimentally investigate turbulent motions in a countless number of applications [119–121]. Some hardware improvements, such as the development of double-oscillator Nd:YAG lasers and specific digital cameras and the growth of computational power (essential to cope with the burdensome image processing), but also the progresses on the theoretical and the "software" sides, have conclusively contributed to such a success.

So far, several works have used PIV for the analysis of turbulent thermal convection [45, 61, 64, 95, 122–125], however, as aforementioned, rare are the cases in which three-dimensional PIV or particle tracking velocimetry (PTV) have been applied [116, 126], although these techniques have already reached a fairly good level of development [127].

This chapter presents the working principles of tomographic PIV, with the goal of clarifying which are the inherent difficulties in applying such a technique to the study of RB convection.

## 2.1. Fundamentals of technique

Tomographic PIV (T-PIV) [128, 129] is an optical technique that allows for instantaneous three-components velocity measurements in multiple points of a flow field by exploiting the scattering of light by small particles that are disseminated in the fluid and follow the flow with sufficiently high accuracy. Specifically, stroboscopic light sources (typically laser or LED) are used to illuminate the seeding particles at two consecutive instants separated by a small time delay $\Delta t$ and the scattered light is recorded by multiple cameras from independent viewing directions. A common setup for T-PIV measurements is shown in figure 2.1, where, in addition, the various steps of the T-PIV processing are outlined.

The acquired images ("projections") are analyzed to reconstruct the 3D distribu-

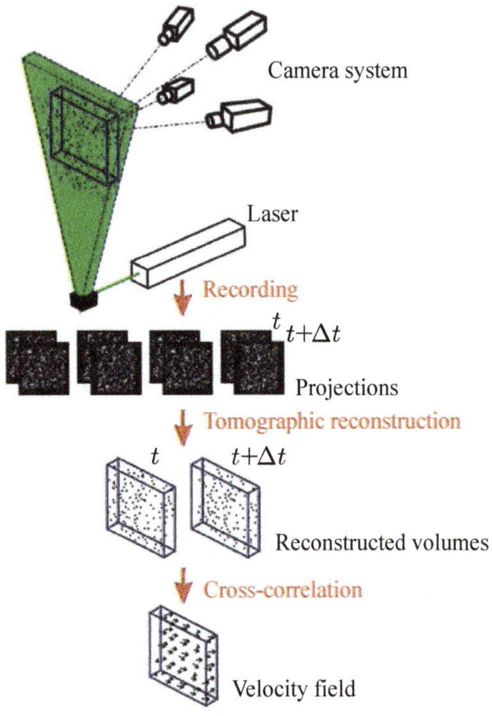


Figure 2.1: Working principle of tomographic PIV. Reproduced from [128].

tion of the light intensity in the investigated volume of fluid by means of tomographic algorithms. After reconstruction, the displacement field between the instants $t$ and $t+\Delta t$ is determined via statistical techniques. For this purpose, the two reconstructed volumes are divided in smaller volumes, called "interrogation volumes" (or "interrogation window"), and 3D cross-correlation between corresponding interrogation volumes yields a displacement vector s for each of these volume pairs. The vector s is representative of the displacement averaged over the time interval $\Delta t$ and over the interrogation volume. A second-order estimate of the velocity vector u in the center of the interrogation volume at the time $t + \Delta t/2$ is obtained simply as $u \approx s/\Delta t$. In such a way, the velocity field is sampled in several points (the centroids of the interrogation volumes, typically arranged on a regular Cartesian grid) within the whole 3D domain.

Indeed, the above described process is comprised of some additional steps that play a crucial role for the efficiency and speed of the process itself. Firstly, for the 3D reconstruction to be accurate, a calibration of the camera system is needed. This usually consists of two stages, a target-based calibration, preceding the recording step and based on capturing images of a target swept through the measurement volume, and a self-calibration [130–132], based on analysis of the acquired particle images. Secondly, in order to both enhance the reconstruction quality and speed-up the involved computations (which may be very extensive for large volumes), particle images are pre-processed. Finally, since the determination of velocity vectors by virtue of cross-correlation does not come without errors, spurious vectors have to be detected and conveniently replaced before further post-processing [133]. Indeed, the latter step is necessary also when the vector field is determined by a multi-resolution and multi-grid approach based on iterative deformation of the interrogation volumes.

In the next sections, the various steps of the T-PIV process are briefly discussed in more detail, thus highlighting the relevant aspects, and also the practical problems, connected with each of them.

## 2.2. Seeding particles

THE selection of seeding particles is among the most critical aspects of the T-PIV setup, especially in experiments on thermal convection.

PIV is an *indirect* anemometric technique, in the sense that the output of the measurement is not the velocity of the fluid under investigation, but indeed that of the tracer particles. This requires, in order to extract reliable information on the flow dynamics, that the particles faithfully follow the bulk motion of the fluid. In addition, seeding particles have to satisfy two further requirements: scatter light of intensity sufficient to form clear images and have an almost uniform size. In the following, focus on the implications of each of these requirements is given. More space is indeed spent for discussion of the first requirement (fidelity of seeding particles as motion tracers), since this aspect may become critical in the investigations on thermal convection.

The equation of motion for a spherical particle suspended in a homogeneous field is the equation of Basset, Boussinesq and Oseen [134, 135]:

$$\mathrm{d}_t \mathbf{U}_p = \frac{\mathbf{U} - \mathbf{U}_p}{\tau_p} + \frac{\rho_p - \rho}{\rho_p}\mathbf{g} + \mathbf{F} \tag{2.1}$$

where $\tau_p$ is the particle relaxation time given by:

$$\tau_p = \frac{\rho_p\, d_p{}^2}{18\mu} \tag{2.2}$$

whereas $\mathrm{d}_t$ denotes the total derivative with respect to time, $\mathbf{U}$ and $\mathbf{U}_p$ are the velocities and $\rho$ and $\rho_p$ are the densities of the fluid and the particle, respectively, $\mu$ is the fluid dynamic viscosity, $d_p$ is the particle diameter and $\mathbf{F}$ is an additional force per unit particle mass. Note that the second term in equation (2.1) is the viscous resistance expressed according to the Stokes' law, which is valid when the particle Reynolds number $Re'_p = \rho\Delta U d_p/\mu$ based on the characteristic slip velocity $\Delta U$, is lower than unity. The latter condition is usually satisfied in thermal convection since seeding particles have typically micrometric size, i.e. $d_p \sim O(10^{-6})$. The third term represents the buoyancy acceleration. As concerns the additional force term, some effects included in it are the "virtual mass" force, related to the inertia of the fluid mass attached to the particle, the forces related to pressure gradients or velocity gradients in the bulk flow and the contribution to resistance due to the unsteadiness of the bulk flow.

It is useful to non-dimensionalize equation (2.1). In this regard, supposing that the particle velocity lag $\mathbf{U} - \mathbf{U}_p$ is small (i.e., the particle is a good tracer), it is possible to assume an unique characteristic velocity scale $U$ for the fluid and the particle velocities. If $L$ is the characteristic length scale of the flow, then the characteristic time scale is $\tau_f = L/U$. Still in the hypothesis that the particle is a good tracer, it has also to follow faithfully the fluid acceleration; therefore, the time derivative $\mathrm{d}_t \mathbf{U}_p$ scales as $U/\tau_f$. Moreover, it could be demonstrated that $\mathbf{F}$ scales as $(\rho/\rho_p)(U/\tau_f)$. With the above premises, non-dimensionalization yields:

$$St_p \mathrm{d}_t \mathbf{u}_p = \mathbf{u} - \mathbf{u}_p + \frac{Ar_p}{Re_p}\hat{\mathbf{g}} + St_p \frac{\rho}{\rho_p}\mathbf{f} \qquad (2.3)$$

where the non-dimensional quantities are denoted by the lower-case variants of the same symbols used for the dimensional counterparts and four more dimensionless parameters appear, namely the ratio of the fluid density to the particle density $\rho/\rho_p$, the particle Reynolds number $Re_p = \rho U d_p/\mu$, the particle Stokes number $St_p$ and the particle Archimedes number $Ar_p$, with the latter two defined as follows:

$$St_p = \frac{\tau_p}{\tau_f} = \frac{\tau_p U}{L} \qquad (2.4)$$

$$Ar_p = \frac{g\, d_p^3\, \rho\,(\rho_p - \rho)}{18\mu^2}. \qquad (2.5)$$

Note that the Archimedes number has in general a sign, being related to the difference of densities between the fluid and the particle.

In equation (2.3), the terms $\mathbf{u}$ and $\mathbf{u}_p$ are unit order, thus the condition of zero velocity lag ($\mathbf{u} = \mathbf{u}_p$) requires that all the remaining terms are sufficiently small. Ultimately, this results in the conditions $St_p \ll 1$ and $Ar_p \approx 0$. The first condition $St_p \ll 1$ reads also as $\tau_p \ll \tau_f$ and has a clear physical meaning: the characteristic response time of the particle (i.e, its relaxation time) has to be sufficiently smaller than the characteristic times of the fluid flow in such a way that the particle can adjust rapidly to any (temporal) variation in the flow itself. The second condition implies neutral buoyancy of the tracer. It should be also noted that in order that the effects of the buoyancy forces are negligible, $Re_p$ does not have to be too small.

So far, the fluid and particle properties have been implicitly supposed to be constant. Nevertheless, in thermal convection the variation of density with temperature cannot be neglected, since it represents the driving force of the motion itself. Within the OB approximation, introducing the relationship (1.1) in equation (2.3), a second Archimedes number can be defined:

$$Ar_p' = \frac{g\, d_p^3\, \rho_0^2\, \beta\theta\Delta}{18\mu^2} \qquad (2.6)$$

where $\rho_0$ now denotes the mean density of the fluid and $\theta_\Delta$ is the temperature deviation from the mean temperature $T_0$. Thus, in this case, the negligibility of the buoyancy forces imposes the additional constraint $Ar'_p \ll 1$.

Obviously, such a condition, along with the previous ones, has to be satisfied locally in any region of the investigated fluid. A more careful analysis reveals that T-PIV (and, generally, PIV) measurements in the kinetic boundary layers forming on the top and the bottom plates of the convection cell is indeed a very hard task. In fact, in the kinetic boundary layer fluid velocities decrease, while the absolute value of $\theta$ increases; therefore, relatively small value of $Re_p$ and large value of $Ar'_p$ are indeed expected, and the buoyancy forces, weighted by the coefficient $Ar'_p/Re_p$ in equation (2.3), may become considerably important. Several investigations have shown that such an effect leads to particle sedimentation on the top and bottom plates [136–140]. Although Okada *et al.* [137] observed, through flow visualization, that the fluid motion during particle sedimentation is almost the same as that of thermal convection without particles, Joshi *et al.* [140] pointed out that the particle deposit constitutes a porous layer with non-ideal thermal properties (and, in particular, low thermal conductivity), which not only decreases the heat flux, as shown by their work, but is also expected to influence the flow dynamics, similarly to non-ideal thermal sources (see section 1.6).

Of paramount importance for particle sedimentation is the particle size distribution. In fact, suspended particles tend to coalesce for their natural tendency to minimize the large specific surface area and excess surface energy [141]. A wide particle size distribution facilitates the formation of particle clusters, which, by virtue of their larger diameters, settle out under the effect of the buoyancy forces. Even when the aggregates remain in the bulk, they do not have the properties of a good tracer because of their long relaxation times.

Following the above discussion, it is evident that good dynamical properties of the seeding particles are ensured by a small diameter $d_p$, a uniform particle size distribution and a density matching as much as possible the fluid density. However, the selection of the particle size is generally a compromise between the dynamical and optical properties of the particles [134]. In fact, the scattering power of small spherical particles is proportional to between $d_p^2$ and $d_p^4$, depending on the light wavelength [142]. On the other hand, the scattering efficiency strongly depends on the ratio of the refractive index of the particles to that of the fluid, according to the Mie's scattering theory [143].

In consideration of all the above factors, seeding particles typically employed in water experiments are large hollow glass spheres and polymer particles with diameters ranging from $10\,\mu$m up to $500\,\mu$m, while air flows are typically seeded with oil droplets or DEHS particles generated by means of a Laskin nozzle with mean diameters of $1\,\mu$m. In fact, in air a perfect density matching between the working fluid and the seeding material is never accomplished and, thus, smaller diameters are generally required to limit the effects of the buoyancy forces; however, this also reduces significantly the amount of scattered light as well. Recently, the use of helium-filled soap bubbles has been proposed as a feasible solution to reach considerably low relaxation times ($\approx 11\,\mu$m) with a brightness $10^4$ times greater than that of standard oil aerosol [144].

## 2.3. Optical system and imaging

T HE two main components of a PIV optical system are the light source and the digital cameras. Additional elements, such as lenses, mirrors and filters, are used, on one side, to shape the light beam in an appropriate pattern, on the other side, to enhance the particle imaging, thus obtaining high contrast (or better, signal-to-noise ratio) images.

In most PIV applications, the high-energy stroboscopic illumination of the flow field is provided by dual-cavity pulsed lasers. These lasers use Q-switching to obtain energetic short pulses with a duration of few nanoseconds and a variable energy, depending on the repetition rate. High pulse energy is achievable only at low or moderate frequencies and thus available for low-speed applications. Nd:YAG lasers are used in these cases, with typical maximum repetition rates of 10 Hz to 15 Hz and energy pulses of 300 mJ. High-speed setups comprise Nd:YLF lasers, which, while operating at frequencies between 0.01 kHz to 10 kHz, are characterized by a considerably lower pulse energy (15 mJ to 30 mJ). The wavelength of the emitted light is 532 nm for the Nd:YAG lasers and 526 nm for the Nd:YLF ones.

In time-resolved tomographic experiments, the low pulse energy of Nd:YLF lasers can preclude investigation of large volumes, as large as the compact convection cell of the present experiments. As a matter of fact, the intensity of the scattered light drops off as the size of the illuminated volume increases. In air experiments, this is an even more limiting factor due to the small size of the employed seeding particles. To face up with the insufficiency of the light energy density, some expedients, like multi-pass light amplification [145, 146], can be used. However, in this regard, it is worth noting that, since thermal convection features very low velocities and, thus, long characteristic times, the repetition rates of the high-energy Nd:YAG lasers are suitable for the investigation of such a phenomenon.

Common PIV cameras comprise CCD (charge-coupled device) and CMOS (complementary metal-oxide-semiconductor) sensors. CCD sensors typically offer higher dynamic range and higher resolution, but they are also slower, more expensive and more power-consuming. CMOS sensors allow for repetition rates up to few thousands per second and have also a better anti-blooming behavior. The more recent sCMOS (scientific CMOS) technology combines the advantages of modern CCD and CMOS sensors to provide high performance. Typical sensor sizes vary between 1 to 16 millions of pixels.

Illumination and cameras are the first ingredients for an optimal particle imaging, but certainly not the only ones involved. High accuracy in measurements requires particles to be in focus throughout the illuminated region and to have images on the sensor plane with an appropriate size (in pixel units). For this purpose, cameras have to be equipped with appropriate lenses.

The particle image is determined by both geometric and diffraction effects, with the former determined essentially by optical magnification and the latter influenced also by the lens aperture area and the light wavelength. Specifically, the diameter of the particle on the sensor is given by [142]:

$$d_\tau = \sqrt{d_{\text{geom}}^2 + d_{\text{diff}}^2} = \sqrt{(Md_p)^2 + [2.44\,\lambda\,f_\#(1+M)^2]^2} \qquad (2.7)$$

where $d_{\text{geom}}$ and $d_{\text{diff}}$ are the contributions related to the geometric and diffraction effects, respectively, $M$ is the magnification factor (equal to the ratio $d_{\text{geom}}/d_p$), $\lambda$ is the light wavelength and $f_\#$ is the $f$-number (ratio of the lens focal length to the aperture diameter). In most cases, $d_{\text{diff}} \gg d_{\text{geom}}$ and $d_\tau \sim f_\#$. Typically, $d_\tau$ is expressed in pixel units and $d_\tau^* = d_\tau/d_{\text{pix}}$, with $d_{\text{pix}}$ being the pixel size in physical units. As a rule of thumb, $d_\tau^*$ should be greater than 2 and lower than 5, leading values smaller than 2 to problems of "peak locking", a bias error in the displacement estimation [147, 148], and values larger than 5 to higher uncertainty in the same computation.

As regards the condition of good focus, the extent of the in-focus region is referred to as "depth of field" and is defined as the distance along the optical axis over which there does not occur a significant blurring of the particle image. Such a distance is indeed related to the diffraction diameter by the equation [119]:

$$\Delta_{\text{field}} = 2f_\# d_{\text{diff}} \left(\frac{M+1}{M^2}\right) = 4.88 f_\#^2 \lambda \left(\frac{M+1}{M}\right)^2. \qquad (2.8)$$

In the case of volumetric measurements, the required depth of field can be considerably large and such requirement can be satisfied only by increasing $f_\#$. Moreover, in some arrangements, cameras are tilted with respect to the midplane of the measurement volume and this causes a reduction of the effective depth of field. This difficulty is overcome by using specific lens-tilt adapters that make the plane of focus parallel to the imaged plane (Scheimpflug condition). Finally, it should be noted that the increase in $f_\#$ leads to a decrease in the recorded light intensity, as a consequence of the decrease of the aperture area.

## 2.4. Camera calibration

CALIBRATION of the camera system is a crucial step in T-PIV process, since the accuracy in tomographic reconstruction is strictly related to the accuracy of the mapping function. The aim of the calibration is to establish a relationship between the 3D world coordinates $\mathbf{x}$ of a point in the physical space and the 2D image coordinates $\mathbf{X}_c$ of its projection on the camera sensor. This projection transformation is defined in terms of a function $\mathbf{m}_c$ (*mapping function*), that depends on a certain number of constants $\mathbf{p}_c$ (*calibration parameters*); the output of the calibration is essentially an estimate of the calibration parameters for each camera of the optical system.

The mapping function identifies a vector correspondence $\mathbf{X}_c = \mathbf{m}_c(\mathbf{x}; \mathbf{p}_c)$ which is not one-to-one, since the same location $\mathbf{X}_c$ on the sensor plane may correspond to

Figure 2.2: Schematic outline of the volume self-calibration procedure for a system of four cameras. Reproduced from [131].

multiple locations $\mathbf{x}_c$ in the physical space. The set of points in the physical space that have the same image on the camera sensor constitutes a *line-of-sight* (LOS). As explained below, the determination of the LOSs of the pixels of the camera sensor is the first step in tomographic reconstruction strategies.

Calibration consists in identifying a sufficient number (generally much greater than the number of calibration parameters) of coordinate pairs $(\mathbf{x}, \mathbf{X}_c)$, in such a way that it is then possible to estimate $\mathbf{p}_c$ by means of optimization algorithms. This process involves typically two steps: the first (*target-based calibration*) is based on recording images of a calibration target, the second (*volume self-calibration*) relies on recording images of the particle themselves.

As regards the target-based calibration, the target is generally made of high contrast markers (dots or crosses) arranged in a regular grid, which act as control points and also enable to identify the coordinate directions of the world reference frame. Images of the target positioned at different known locations within the measurement volume are acquired. For each of these images, the 3D world coordinates $\mathbf{x}$ of the control markers are precisely known, whereas the 2D image coordinates $\mathbf{X}_c$ are determined by template-matching techniques using cross-correlation or detection methods based on threshold and geometric computation of the centroids.

In a target-based calibration there might be some factors that could compromise the accuracy in the identification of the calibration parameters, namely manufacturing imperfections, residual errors in determining the markers' image positions and misalignment in target positioning. Volume self-calibration [130, 131] allows for a further adjustment of the calibration parameters, causing the residual errors to be reduced from 0.5–1 voxels, which are typical values for a target-based calibration, to less than 0.1 voxels.

The working principle of volume self-calibration is schematically represented in figure 2.2 for a system of four cameras. The basic idea of the method is to triangulate the 3D positions of the particles in the illumination volume starting from their 2D projections and determine the disparities between such projections and the back-projections of the triangulated positions. To do this, as a first step, it is necessary to match the particle images on the four cameras that correspond to the same particle in the physical space. Starting from a particle image on the first camera, the projection of its LOS on the second camera's sensor plane is computed. Around this line, possible matching particle images are searched within a rectangular region defined by a maximum search radius (indicated as $\varepsilon_s$ in the figure). For each match found, a guess 3D position of the particle in the physical space is determined via least-squares triangulation (at this stage, using only two particle images). This position is then projected on the next camera's sensor plane and, once again, matching particle images are searched in the neighbor of this projection (a circular area of radius $\varepsilon_s$). For each new match, the 3D position of the particle is again triangulated and projected onto the last camera to find further matches. If a match between all the four cameras is found, the corresponding triangulated particle is used to compute a disparity vector for each camera. Disparities are then represented as Gaussian or parabolic functions in a histogram map (*disparity map*) and summed up over sub-volumes of the entire measurement domain. The peaks in the disparity maps corresponding to the different sub-volumes constitute residual triangulation errors that can be used to correct the calibration parameters.

In T-PIV applications, the most commonly used camera models are the pinhole camera model [149, 150] and the Soloff's model [151]. Based on perspective laws, the pinhole camera model treats the LOSs as straight lines, whereas lens non-linearity is included via second-order spherical distortion terms. The Soloff's model uses a polynomial function with second order in the depth-of-field direction (namely $z$) and third order in the two orthogonal directions ($x$ and $y$). Indeed, the Soloff's model is generally used for illuminated volume with limited thickness in the $z$-direction. For larger volumes, multi-plane polynomial calibration models are used. Such models use independent third-order polynomials on selected planes at different $z$ location within the measurement volume to perform third-order polynomial interpolation in the $x$ and $y$ directions and piecewise linear interpolation along the $z$-direction.

In the comparison with polynamial functions, the main advantage of the pinhole camera model is its simplicity and the physical relevance of the calibration parameters involved. However, this model does not include the effect of distortions along the LOSs (except the localized lens distortions). Optical misalignment, thermal deformation and refraction by optical windows or fluid interfaces are some of the elements not accounted for by the model and that might lead to potentially large calibration errors. When one or several of these effects become important, polynomial mapping functions generally offer higher performance. In the experiments of the present work, the optical deformation caused by the curvature of the cylinder sidewall compromises the application of the pinhole camera model. However, to retain the inherent simplicity of this model, a consistent modification of the same is introduced in chapter 4. The advantages in the comparison with polynomial models and limits of application of this novel model are also extensively discussed.

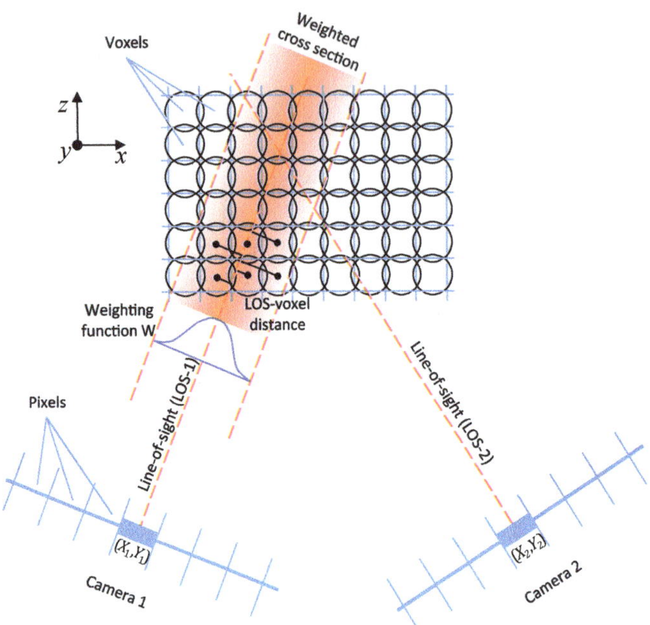


Figure 2.3: Schematic representation of the voxel discretization of the measurement volume and imaging model adopted in tomographic reconstruction. The weighting function for the pixel denoted by 1 is non-zero only in the red-shadowed region; thus, only the voxels falling in this region contribute to the light intensity of the pixel. Reproduced from [129].

## 2.5. Tomographic reconstruction

### 2.5.1. Statement of reconstruction problem

TOMOGRAPHIC reconstruction aims at determining the 3D distribution of the scattered light intensity $E(\mathbf{x})$ starting from its projections on the cameras, which constitute a set of 2D function $I_c(\mathbf{X})$. Since the investigated fluid is typically transparent, the projections $I_c(\mathbf{X})$ can be regarded as integrals of $E(\mathbf{x})$ along the viewing directions. More specifically, the functions $I_c(\mathbf{X})$ are known in a discrete form, i.e. the light intensities recorded by the cameras' pixels $I_i^{(c)}$; likewise, $E(\mathbf{x})$ is discretized by dividing the illuminated volume in smaller portions (*voxels*) where the light intensity is supposed to be constant and equal to $E_j$. The relationship between the 3D intensity values $E_j$ and the 2D intensity values $I_i^{(c)}$ can be approximated by the following linear equations:

$$\sum_{j=1}^{m} w_{ij}^{(c)} E_j = I_i^{(c)} \quad \text{with } i = 1, \ldots, l, \quad c = 1, \ldots, N \tag{2.9}$$

where $i$, $j$ and $c$ are the indexes for the $i$-th pixel, the $j$-th voxel and the $c$-th camera, $l$ is the number of voxels, $n$ the number of pixels, $N$ the number of cameras and

$w_{ij}^{(c)}$ is a weighting coefficient, which expresses the contribution of the light intensity over the $j$-th voxel to the light intensity over the $i$-th pixel of the $c$-th camera. The coefficient $w_{ij}^{(c)}$ is typically calculated as the fraction of the volume of the $j$-th voxel intersecting the LOS of the $i$-th pixel of the $c$-th camera and, thus, it varies between 0 and 1. For simplicity and speed of computation, the voxel is represented as a sphere, while the LOS is schematized as a cylinder, as shown in figure 2.3. Since only a small percentage of voxels contributes to the pixel light intensity, a large number of weighting coefficients is indeed equal to zero.

Equations (2.9) are a system of $l \times N$ equations in $m$ unknowns, which can be concisely written as $\mathbf{WE} = \mathbf{I}$. Since $m$ is usually much larger than $l \times N$, the system is undetermined and its solution is not feasible by matrix inversion; on the contrary, iterative approximate methods have to be used. In the next subsection a brief overview of the *algebraic* methods [128, 152–154] is given.

### 2.5.2. Algebraic techniques

ALGEBRAIC techniques are divided in two main categories: *additive* and *multiplicative*. The standard methods belonging to these two classes are ART (algebraic reconstruction technique) [155] and MART (multiplicative algebraic reconstruction technique) [156]. The update equations from the $k$-th iteration to the $(k + 1)$-th one for these two methods are:

$$\text{ART:} \quad E_j^{(k+1)} = E_j^{(k)} + \bar{\mu} \frac{I_i^{(c)} - \sum_j w_{ij}^{(c)} E_j^{(k)}}{\sum_j w_{ij}^{(c)}} w_{ij}^{(c)} \tag{2.10}$$

$$\text{MART:} \quad E_j^{(k+1)} = E_j^{(k)} \left( \frac{I_i^{(c)}}{\sum_j w_{ij}^{(c)} E_j^{(k)}} \right)^{\bar{\mu} w_{ij}^{(c)}} \tag{2.11}$$

where $\bar{\mu}$ is a relaxation coefficient that determines the stability of the method (the stability criterion for MART is that $0 < \bar{\mu} < 2$).

The above equations show that in ART the update term depends on the difference between the projection $I_i^{(c)}$ and the 3D light field back-projection $\sum_j w_{ij}^{(c)} E_j^{(k)}$, which is in fact the residual of $\mathbf{WE} - \mathbf{I}$ equation (2.9); thus, the convergence of the method implies that this residual is zero. Vice versa, the update term of MART is related to the ratio of the projection $I_i^{(c)}$ to the 3D light field back-projection and convergence implies that this ratio is equal to the unity. By virtue of such a property, MART works as an AND operator, in the sense that the voxel intensity $E_j$ is non-zero only if the intensities $I_i^{(c)}$ of all the pixels that "view the voxel" ($w_{ij}^{(c)} > 0$) are non-zero. This is, in general, the reason why multiplicative methods generally work better than additive ones. A typical number of 5 MART iterations is, in most cases, sufficient to reach convergence [129].

In literature, several variants of the ART and MART algorithms have been formulated and their performance investigated (see, for instance, [157, 158]). A review of such methods is beyond the scope and the focus of this dissertation. However, it is worth noting that, when no further information than that contained in the projections $I_i^{(c)}$ is exploited, MART has proved to be the best choice in the comparison with other methods (which, although less expensive than MART, require a larger number of iterations to achieve equivalent reconstruction quality). Beyond MART, a significant improvement in the reconstruction quality can be obtained by the so-called "motion tracking-enhanced" (MTE) methods, which are presented in section 2.5.4.

## 2.5.3. Ghost particles

$\mathrm{F}$OR the discussion to follow, it is useful to mention that tomographic reconstruction by conventional algebraic methods is affected by unavoidable errors that affect the uncertainty of the computed displacement. As noticed by Novara *et al.* [159], these errors are essentially of three kinds: discretization errors, due to reconstructed particles smaller than 3 voxels, elongation of the particles along the depth direction due to small angles between adjacent cameras and formation of *ghost particles*. Ghost particles are regions of non-zero light intensity formed at the intersection of LOSs where no actual particle is indeed located in the physical space. Figure 2.4a shows schematically the emergence of ghost particles in the case of a 2-camera setup imaging two actual particles.

Ghost particles are the dominant source of error in T-PIV measurements, especially at high seeding densities. Indeed, at low to moderate seeding densities, ghost particles exhibit lower intensity than actual particles [128, 160] and, in principle, they might be detected and removed from the reconstructed volumes. In the displacement estimation by cross-correlation, the effect of ghost particles is twofold: on one side, ghost particles constitute a noisy background that affects the cross-correlation map, potentially reducing the peak intensity and increasing uncertainty in its detection; on the other side, pairing of ghost particles between consecutive snapshots can result in a modulation of the displacement field. Figure 2.4b reports a simplified case in which a ghost particle outlives the displacement of the actual particles from the LOSs of which it has been generated and is found also in the following snapshot, whereas, in the case drawn in figure 2.4c, the ghost particle is present only in one exposure. Although in the first case (figure 2.4b) the ghost particle displacement is coincident with the mean displacement of the "parent" actual particles, the ghost particles may be located very far from the parent particles and thus they can spread the actual velocity information out over the entire volume (and also outside of it). This results in a velocity modulation and subsequent reduction of the velocity gradient.

In the case of a $N$-camera system, an approximate estimate of the ratio of the number of ghost particles $N_g$ to the number of true particles $N_p$ is given by [162]:

$$\frac{N_g}{N_p} = ppp^{N-1} A_p^{N-2} d_\tau^* \frac{M \Delta_{\text{field}}}{d_{\text{pix}}} = \frac{4}{\pi d_\tau} N_S^{N-1} M \Delta_{\text{field}} \qquad (2.12)$$

where $ppp$ (particle per pixel area) is the particle image density, while $N_S$ is the so-called *source density* given by the product of $ppp$ and the average particle area in pixels (therefore, $N_S$ equals the fraction of the area sensor that is actually illuminated). Since the ghost particles are artifacts of the tomographic reconstruction, the ratio

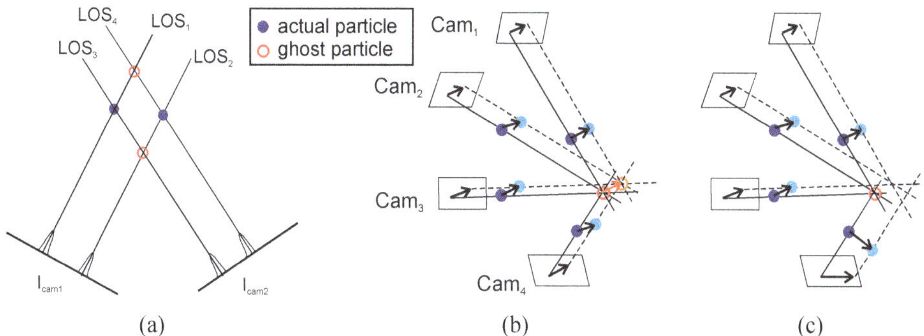


Figure 2.4: Ghost particles: (a) formation of ghost particles in tomographic reconstruction with a 2-camera setup; (b) example of ghost particle surviving through two consecutive snapshots and contributing to the displacement modulation; (c) example of ghost particle passing away through two consecutive snapshots. Reproduced from [161].

$N_p/N_g$ is an indicator of the signal-to-noise (SN) ratio. Therefore, equation (2.12) suggests that an increase of the seeding density leads to a deterioration of the SN ratio, due to the increase in the number of ghost particles. The same effect is brought about by an increase of the depth-of-field and a decrease of the particle image diameter at constant $N_S$. In fact, the latter effect, which could seem counterintuitive, can be explained considering that a decrease in $d_\tau$ at constant $N_S$ implies an increase in the number of true particles, which corresponds to a larger increase in the number of LOS intersections (and, thus, of the ghost particles).

A direct corollary of the above considerations is that there exists an optimal value of the source density $N_S$ for given values of the depth-of-field and the imaging properties of the optical system (included the particle size). It is also noted that the spatial resolution of T-PIV measurements is related to the seeding density, since the requirement for a reliable displacement estimate is that at least 5–7 particles fall in the interrogation window [163]. Typical suggested values of $N_S$ are around 0.2 (the total area occupied by particles in the recording is the 20% of the total imaging area), which corresponds, generally, to $ppp \approx 0.05$ [129]. Beyond this threshold, the detrimental effects of the ghost particles lead to unacceptable levels of uncertainty in velocity field estimation. Nevertheless, strategies capable of reducing the number of the ghosts, like the MTE methods, can allow for accurate measurements up to $ppp \approx 0.1$. Therefore, such methods are essential in investigation of large-scale volumes (like that of the present experiments), which are characterized by large values of $\Delta_{\text{field}}$.

### 2.5.4. Motion tracking-enhanced techniques

MTE methods are a recently developed class of reconstruction techniques that exploit the coherence of particle motion over two or more frames to enhance the tomographic reconstruction by reducing the number of ghost particles. The basic idea of this approach can be explained with the aid of figure 2.4c: while actual particles always occur in pair between a pair of snapshots separated by a small time

$\Delta t$, ghost particles sometimes do not, and such a feature can be exploited to detect and remove them or at least to smooth out their intensity in the reconstruction process.

The first implementation of the MTE approach is due to Novara *et al.* [159]. In their algorithm (MTE-MART), as a first step, a pair of snapshots, namely the 3D light fields $\mathbf{E}_n$ and $\mathbf{E}_{n+1}$ (taken at the times $t$ and $t + \Delta t$, respectively), is reconstructed by MART and the corresponding displacement field $\mathbf{u}_n$ is computed via cross-correlation. Secondly, the snapshot $\mathbf{E}_n$ is deformed according to this displacement field to obtain an *a-posteriori* estimation $\mathbf{E}'_{n+1}$ of the 3D light field at the time $t + \Delta t$. Similarly, the backward (time) projection of the snapshot $\mathbf{E}_{n+1}$ provides an *a-posteriori* estimation $\mathbf{E}'_n$. Then, the fields $\mathbf{E}''_n = \frac{1}{2}(\mathbf{E}_n + \mathbf{E}'_n)$ and $\mathbf{E}''_{n+1} = \frac{1}{2}(\mathbf{E}_{n+1} + \mathbf{E}'_{n+1})$ are computed. Finally, the latter are used as first guess for a new MART reconstruction. The above described steps can be repeated until the result exhibits convergence.

The strength of this method lies in performing, at each iteration, the arithmetic mean of one snapshot with its *a-posteriori* estimate. In fact, in such operation the intensity of particles that do not match between the two snapshots is reduced roughly by a factor of 2; as the iterations go on, these potential ghost particles are therefore smoothed out and their initial intensity is redistributed among the actual particles, which are then reconstructed more accurately.

Novara *et al.* [159] also demonstrated, through numerical simulations, that the maximum seeding density that can be treated with the MTE-MART algorithm is tripled with respect to the conventional MART. However, the method is computationally expensive, given its iterative nature.

A recent development of MTE-MART is the sequential MTE-MART (SMTE-MART) algorithm [164], which is suitable for time-resolved measurements. SMTE-MART relies on a time-marching estimation of the 3D light field to obtain an enhanced first guess given as input to MART. Differently from MTE-MART, the displacement field $\mathbf{u}_n \equiv \mathbf{u}'_{n+1}$ is used to forward-project the field $\mathbf{E}_{n+1}$ and estimate the initial guess field $\mathbf{E}^G_{n+2}$ for the MART reconstruction. This is now an *a-priori* estimation of the light field at the successive time instant. In comparison with MTE-MART, such an approach has two key advantages. First, the only iterations required are those related to the MART reconstruction, with a considerable saving of computational time; indeed, the enhanced guess potentially allows even for a reduced number of the MART iterations. Second, as the process goes ahead, eventually all the ghost particle are removed from the reconstructed volume since, in the long term, each ghost particle will experience soon or later an evolution similar to that represented in figure 2.4c.

Among all the state-of-art reconstruction methods based on voxel discretization of the light intensity field, SMTE-SMART is by far the most efficient one, although its applicability is limited to time-resolved measurements. However, other advanced techniques have been developed recently, which relies on a completely different approach, i.e., particle reconstruction by means of triangulation, in a fashion very similar to methods traditionally employed in 4D particle tracking velocimetry (PTV). These techniques are the iterative particle reconstruction (IPR) method proposed first by Wieneke [165] and 'shake-the-box' (STB) which, based on IPR, was developed by

Schanz *et al.* [166] at DLR. Both methods are briefly outlined below.

## 2.5.5. Iterative particle reconstruction and 'shake-the-box'

THREE-DIMENSIONAL PTV is based on detection of the 2D particle image locations and successive triangulation of their 3D positions in the physical space. By pairing of 3D particles between two successive snapshots, it is then possible to determine the velocities of the tracer particles. In its classical implementation, such an approach requires low seeding density for unambiguous identification of particle pairs (in most cases, lower than $0.001\,ppp$) and thus suffers from sparsity. In other words, even though the computed velocity field is not modulated as in T-PIV, it is measured at irregular positions and often interpolation of such scattered data to a regular grid for further analysis is in fact not feasible without smoothing and, thus, modulation of the signal.

IPR [165] has overcome these limits of 3D PTV with an hybrid approach between algebraic reconstruction techniques and triangulation. In IPR, a particle is not defined only by its position in the physical space but also by its geometric and optical properties, i.e., diameter and peak intensity. Iterations are needed for optimizing all these properties. This is done by projecting the 3D particle distribution, obtained by triangulation as in 3D PTV, onto the image planes of the cameras and comparing the back-projections with the original images. The locations, intensities and diameters of the particles are refined by minimizing the norm of the residual image, defined by the difference between the back-projection and the original image. In such an iterative procedure, particles that happen to get intensity or diameter below certain thresholds are identified as ghost particles and removed; this allows for further refinement of the properties of the true particles.

The fundamental advantage of IPR is its ability to operate at seeding densities higher than up to 100 times those of the conventional PTV. However, for a good quality reconstruction, a large number of iterations is typically required with times comparable to those of MART (and slower than sparse implementation of MART [157]).

The STB technique [166, 167] can be regarded as a motion-tracking enhanced variant of IPR that exploits the coherence of particle motion over time to distinguish and remove ghost-particles. As SMTE-MART, STB applies only to time-resolved measurements. The basic idea of this method is to extrapolate particles along their trajectories to predict their positions at a successive time instant. These positions are refined by 'shaking' (an optimization consisting in minimization of the residual images in the same way as in IPR), then new particles are introduced and refined by triangulation/IPR. The latter step is indeed necessary because at any time there are some particles leaving or entering the measurement volume.

In the comparison with T-PIV reconstruction methods, the two main advantages of STB are essentially the drastically (10 to 100 times) lower computational time with respect to the SMTE and the higher accuracy [168]. In principle, determination of the velocity vectors does not suffer from modulation, but only from reconstruction errors (dictated by possible calibration inaccuracies, background reflections and camera noise, etc.), which, however, when measurements are performed in a high-frequency regime, can be significantly reduced by temporal filtering of particle positions, velocities and even accelerations along trajectories.

It is worth noting that, although IPR and STB should be classified as 4D PTV techniques, hybrid methods that combines reconstruction strategies typical of T-PIV with the particle-tracking ones are indeed not only possible, but also potentially susceptible of better performance. Such a combination can be obtained by switching between voxel-representation and particle-based representation of the light intensity field, by either converting the 3D particles into artificial light blobs or extracting them from voxelized volumes.

In this optic, the present dissertation often uses the locution "time-resolved T-PIV" to indicate, without distinction, both cross-correlation-based tomographic 3D PIV and 4D PTV.


## 2.6. Velocity field estimation

E STIMATION of the velocity fields in PIV is commonly based on cross-correlation of interrogation volumes [120, 169, 170], as explained in section 2.1.

The normalized windowed cross-correlation function $\mathcal{C}$ between the two interrogation volumes $W_n$ and $W_{n+1}$ reconstructed from the recordings at time instants separated by the interval $\Delta t$ is given by:

$$\mathcal{C}(\mathbf{u}) = \frac{\sum_{\mathbf{x}} \phi_W^2(\mathbf{x}) \left(W_n(\mathbf{x}) - \mu_n\right) \left(W_{n+1}(\mathbf{x} + \mathbf{u}) - \mu_{n+1}\right)}{\sqrt{\sum_{\mathbf{x}} \phi_W^2(\mathbf{x})(W_n(\mathbf{x}) - \mu_n)^2 \cdot \sum_{\mathbf{x}} \phi_W^2(\mathbf{x})(W_{n+1}(\mathbf{x} + \mathbf{u}) - \mu_{n+1})^2}} \tag{2.13}$$

where $\mathbf{x}$ and $\mathbf{u}$ now denotes vectors of integer indexes and integer displacements (in pixel units), respectively, $\mu_n$ and $\mu_{n+1}$ are the mean intensities over the volumes $W_n$ and $W_{n+1}$ and $\phi_W(\mathbf{x})$ is a weighting window. If $\phi_W(\mathbf{x})$ is constant (top-hat window), equation (2.13) returns the standard normalized cross-correlation. The value $\mathbf{u}$ for which $\mathcal{C}$ exhibits a maximum represents the average displacement of particle within the interrogation volume. Since the correlation map is discrete, in order to achieve sub-voxel accuracy the correlation peak is typically detected by Gaussian fitting [119, 169].

In volumetric measurements, the accuracy in the displacement estimation is mainly influenced by two effects, i.e., the out-of-volume displacement of particles occurring for large values of the displacement itself (compared to the dimension of the interrogation volume) and the presence of velocity gradients within the interrogation volume. These effects are competing because small volumes are desirable to avoid large velocity gradients, but the smaller the volume, the greater the probability that particles fall outside of it owing to the flow displacement. To avoid both these effects, multi-grid iterative window deformation (MGIWD) is generally employed [171]. This basically consists in determining, as a first step, a predictor displacement field on a rather coarse grid, which is used to deform the interrogation volumes accordingly; a gradual grid refinement in the successive steps allows to determine a correction for the predictor by means of cross-correlation of the deformed volumes. In such a way, not only it is possible to recover most of the particle pairs between the two exposures, but also the modulation effect due to velocity gradients is limited by using smaller volumes in the final iterations of the procedure.

Computation of the direct cross-correlation is among the most time-consuming tasks of the T-PIV process and, since it occurs a significant number of times in the MGIWD process, several expedients are used to speed it up. A first acceleration is provided by performing the cross-correlation in the frequency domain using the fast Fourier transform (FFT) and the convolution theorem [120]. This reduces the number of multiplications from $o(N_W^2)$ to $N_W \log(N_W)$ for each linear dimension. Other fundamental guidelines for a significant speed-up of the cross-correlation have been given by Discetti and Astarita [172]. On one side, they suggested to use, in the first iterations of the MGIWD process, FFT cross-correlation on blocks to avoid redundant calculations in case of overlapping interrogation volumes. The blocks have dimensions equal to the selected overlaps and the correlation map in each interrogation volume is obtained by summing up the maps related to the constituent blocks. On the other side, they proposed to determine the correction field in the final iterations by computing direct cross-correlation only for a maximum displacement in each direction of $\pm 1$ voxel (correction is expected to be at a sub-voxel level). This offers the advantage of a smarter and a more accurate computation. In addition, direct cross-correlation can exploit the sparsity of the reconstructed volumes (typically not greater than 90%, since $ppp$ is at most 0.1) for an even more rapid displacement estimation.

# 3

# Experimental and numerical arrangements

## 3.1. Experimental setup

A sketch of the experimental apparatus is shown in figure 3.1. The details of the convection cell and the temperature control systems related to it are given in section 3.1.1. Section 3.1.2 describes the optical system used for the T-PIV measurements. Finally, the features of the rotating system employed are reported in section 3.1.3.

### 3.1.1. Convection cell

THE convection cell consists of a Plexiglas cylinder filled with water. The internal diameter of the cylinder is $D = 74.0\,\text{mm}$, the aspect ratio $\Gamma$ is one half (height equal to twice the internal diameter) and the thickness of the sidewall is $3\,\text{mm}$.

The cylinder is immersed in an octagonal tank, which is also filled with water. The tank is made of a mammoth resin frame obtained by means of the 3D printing technology and Plexiglas windows which are glued to the columns of the frame with silicone sealant. The shape of the tank is not regular and has been designed in such a way that the directions normal to four adjacent windows form angles of $40°$ between each other, while the relative angles between the remaining four windows are $30°$. Moreover, each set of adjacent windows is located at the same distance from the cylinder axis. The above design ensures that the digital cameras can be easily arranged in such a way to have the same spatial resolution and a constant angular spacing. For this purpose, in fact, it is sufficient to position each camera with its optical axis normal to the Plexiglas window at a fixed distance from the latter and in a such a way that the cylinder is centered in the image. The orthogonality between the camera axis and the Plexiglas window ensures minimal refraction distortions of the optical rays through the window itself.

The convection inside the cylinder is induced by heating the bottom and cooling the top; a high-precision thermo-electric control is used to maintain the temperatures of the cell bases constant to $\pm 0.01\,°\text{C}$ throughout the duration of the experiment. The several components of the temperature control systems are shown in figure 3.2.

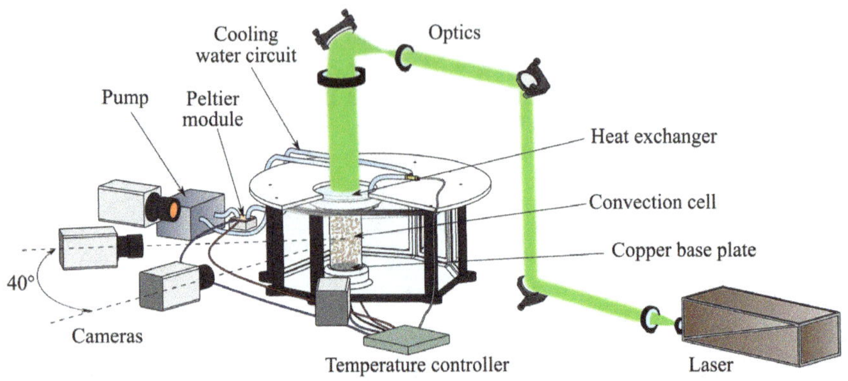



Figure 3.1: Schematic of the experimental apparatus.

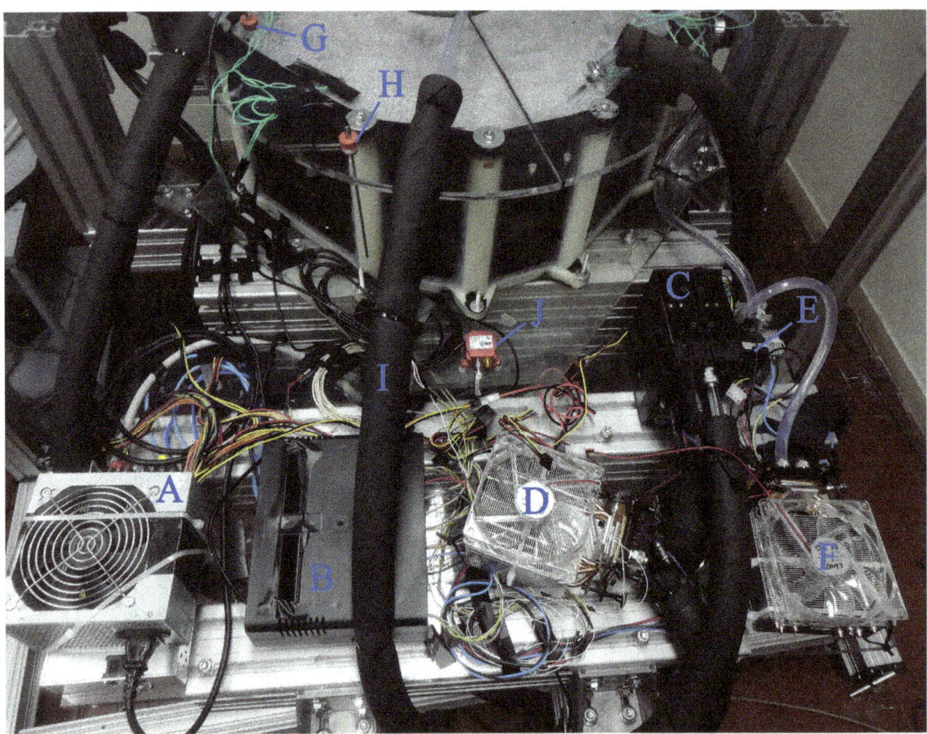



Figure 3.2: The temperature control systems of the experimental apparatus. A: power supplies; B: TEC controllers; C: pump of the water circuit feeding the heat exchanger at the top of the convection cell; D: Peltier device along the circuit of the top cooling system; E: pump of the water circuit for cooling of the tank water; F: Peltier device along the circuit of the tank cooling system; G: immersion thermal probe for measurement of the tank water temperature; H: thermal probe for measurement of the ambient temperature; I: tubes of the water circuits coated with neoprene rubber; J: inertial measurement unit (used in rotating experiments).

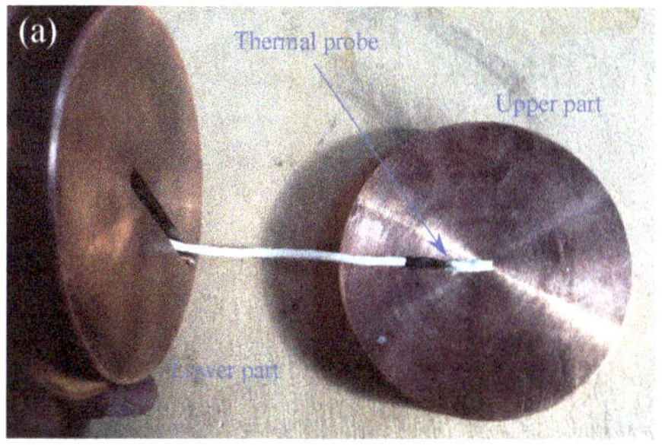


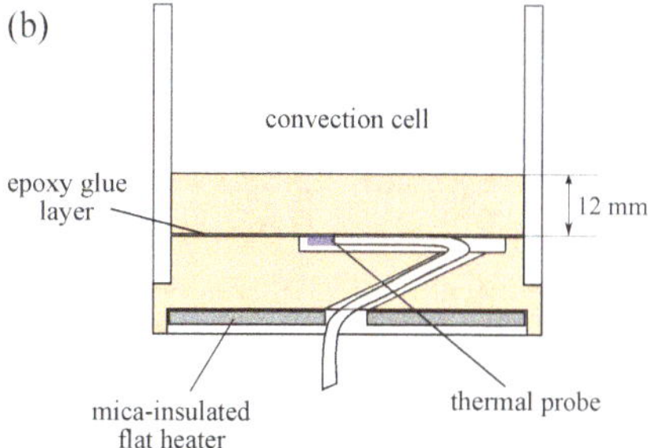


Figure 3.3: Bottom heating system: (a) photograph of the copper insert before gluing the two parts; (b) sketch of the connection between the copper insert and the cylinder.

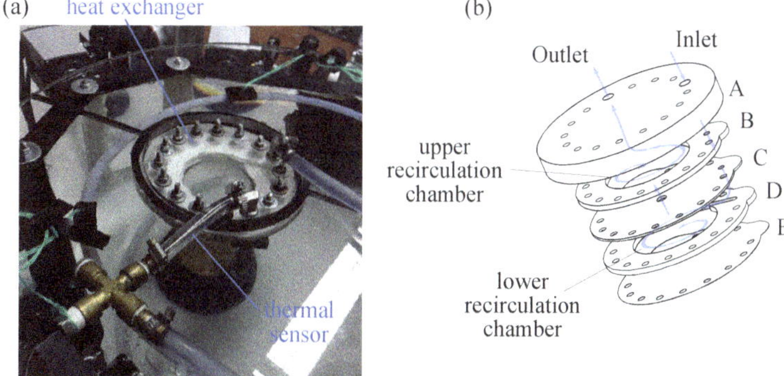


Figure 3.4: Top cooling system: (a) top view with the cylinder in situ; (b) structure of the heat exchanger at the top of the convection cell.

The reader is referred to the caption for a description of the devices shown in the picture.

The bottom heating system (not visible in figure 3.2) consists of an electrolytic copper insert connected to a mica-insulated flat heater, which can provide a heating power up to $100\,\text{W}$. The copper insert is made of two parts in between which a PT100 1/10-DIN RTD sensor (Omega RTD-3-F3105-80-T-1/10) is embedded (figure 3.3). A highly conductive epoxy (Omega OB-101-16) is used to glue this sensor to the upper part, as shown in figure 3.3, and the two parts to each other. The sensor is placed at a depth of $12\,\text{mm}$ below the cell bottom (figure 3.3b); due to the high conductivity of the copper, the temperature drop across this layer is negligible and the sensor measurement provides the actual temperature of the cell bottom. The copper insert is housed in an apposite recess realized on the bottom of the tank. The cylinder is fastened to the copper insert with a slight interference fit; thus, the cylinder sidewall is in contact with the copper insert for a depth of about $20\,\text{mm}$.

The top cooling system consists of a water heat exchanger obtained by assembling a set of Plexiglas layers, as shown in figure 3.4b. These layers comprise two recirculation chambers (the upper chamber consisting of the layers A–C and the lower chamber consisting of the layers C–E) with nearly opposite tangential inlet and outlet, separated by a thin slab with a small central hole (layer C). The various layers are silicone-glued and pressed against each other with the aid of a set of screws and bolts inserted in 14 holes evenly spaced along the periphery of the heat exchanger. This prevents leakages due to the over-pressure of the water pumped into the recirculation chambers. As concerns the pump, we opted for a small size solution and used a model designed for desktop PC water cooling; specifically, the pump is Alphacool VPP655 with a maximum flow rate of $1500 \, \text{L/h}$ and is assembled with the Alphacool Eisfach Bay Res, a reservoir specifically designed for such model of pump (figure 3.2C). The reservoir is fundamental to eliminate bubbles that are entrapped in the circuit in the initial phase of filling. For this purpose, the water pump is initially operated at its maximum rotational speed; after all the bubbles have left the circuit, the pump is driven to the minimum rotational speed and the bay is completely filled and hermetically closed to avoid new air to enter the watercircle.

Before entering the heat exchanger, the water is refrigerated by a Peltier device (figure 3.2D). This device consists of a copper waterblock connected to a Peltier element (Mouser Electronics P/N:490-CP85435) with a maximum heating/cooling power of $118 \, \text{W}$ and a maximum allowable temperature difference between its sides of $\pm 75°$; to avoid overheating of the Peltier element, the latter is connected to a heat sink with fan. It is worth noting however that both the bottom and the top systems are oversized with respect to the thermal power required by the RB convection in the cell and, in the steady operation, the current loads of both the heater and the Peltier element are considerably small. The water tubes of the cooling circuit are thermally insulated by neoprene rubber coating (figure 3.2J) in order to prevent heat losses due to thermal exchange with the surrounding ambient.

The heat exchanger water temperature is measured at the heat exchanger inlet by an immersion PT100 1/10-DIN RTD sensor (Omega P-M-1/10-1/8-6-1/4-P-6), as shown in figure 3.4a. Such temperature differs from the actual temperature of the cylinder top because of the temperature drop across the lower Plexiglas layer of the heat exchanger (layer E in figure 3.4), which is directly in contact with the convective fluid. To reduce such a drop, this layer is just 0.25 mm thick (it consists of a transparent Goodfellow PMMA sheet). The temperature drop for the present experiments reported is expected to be at worst equal to $0.2 \, °\text{C}$. This estimation is based on a one-dimensional thermal analysis assuming that the power heat flowing through the cell is equal to the power output provided by the TEC to the Peltier element; therefore, it is conservative since it does not include heat losses through the circuit. A better estimation can be however obtained from the computation of the heat transfer through the cell; this is done later in chapter 5 based on the results of the numerical simulations carried out in analogous conditions to those of the experimental tests.

As aforementioned, the temperatures of the copper slab and the refrigerating water are continuously monitored and maintained constant by a high-precision thermo-electric controller (TEC) (figure 3.2B). The TEC (Meerstetter Engineering TEC-1123-HV) performs a PID control by adjusting the current inputs to the flat heater and the Peltier element and ensures a temperature stability of $0.01\,^{\circ}\mathrm{C}$, as aforementioned. Since the temperature difference between the top and the bottom plates is $5\,^{\circ}\mathrm{C}$ in the experiments presented here, this corresponds to a percentage stability of the imposed unstable temperature gradient of $0.2\%$.

The early version of the experimental apparatus was not designed to control the temperature of the water in the tank surrounding the convection cell; such a temperature was simply measured by means of an immersion PT100 1/10-DIN RTD probe (Omega P-M-1/10-1/8-6-0-P-6) (figure 3.2G). After the experiment in non-rotating conditions, it was realized that the temperature at the external side of the sidewall has a significant influence on the fluid dynamics of the RB convection, despite the low thermal conductivity of the Plexiglas wall. A more detailed discussion about this point is given in chapter 5. To obviate this problem, a specific temperature control system was designed for the rotating experiments. This system is similar to that employed for the temperature control of the heat exchanger placed at the top of the convection cell; it consists of a water circuit comprising a pump (figure 3.2E), a Peltier cooling device (figure 3.2F) and two loop tubes immersed in the tank: water is withdrawn from the bottom of the tank, refrigerated by the Peltier cell passing through a copper waterblock and pumped back to the tank from the top. This generates a convective motion outside of the cylindrical sample, which allows to maintain the temperature of the external side of its wall constant at the tank bulk temperature. Such a temperature is in fact monitored by means of the thermal probe and controlled by a second TEC of the same kind as that employed for the top and the bottom plates. It is worth noting that the canonical problem of RB convection is defined inside a cylinder with adiabatic lateral wall. Obviously, the requirement of optical accessibility makes it impossible to reproduce such a condition in the present experimental apparatus. However, as shown later, an isothermal condition of the external side of the cylinder sidewall at a temperature equal to the average between the top and the bottom temperatures leads to an evolution exhibiting strong similarities with the canonical case.

## 3.1.2. Imaging system

THE T-PIV system consists of a dual pulse Nd:YAG laser with maximum pulse energy of $200\,\mathrm{mJ}$ and four sCMOS cameras with a resolution of $2560 \times 2160$ pixels. The laser light is shaped into a cylindrical beam that passes through the transparent heat exchanger on the top of the cell and illuminates the entire convection domain. The cylindrical shape is obtained by using an appropriate system of mirrors and lenses, as shown in figure 3.1.

The four cameras are arranged in a planar configuration and the angular spacing chosen in the present experiments is $40^{\circ}$. In order to focus the whole cylinder interior, $28\,\mathrm{mm}$ focal length objectives set at f-number equal to $22$ are used. The short focal length allows to reduce the distance of the cameras from the measurement volume and this is crucial in the rotating experiments in order to limit the inertial forces acting on the cameras themselves. The digital resolution of the cameras is $14$ voxel/mm for the non-rotating experiments and $15$ voxel/mm for the rotating ones.

The seeding particles used in the experiments are orange fluorescent polyethylene microspheres (Cospheric UVPMS-BO-1.00); the average particle diameter is $58\,\mu m$, while the particle density is 1.00 g/cm$^3$, resulting in a relaxation time lower than 1 ms (which is significantly below the turbulent dissipative time scales of the thermal convection at the currently investigated conditions). In some preliminary tests, polyamide Vestosint powder with a nominal density of 1.016 g/cm$^3$ and nominal diameter of $56\,\mu m$ was used, but it was observed that this kind of seeding particles undergoes rapid sedimentation, which makes it unfeasible to perform measurements over a long time, as in the present experimental investigations. The faster sedimentation in the latter case is associated, on one side, with the greater density of the particles, on the other side, with the large particle size distribution (section 2.2). It is worth noting that sedimentation occurs also in the case of polyethylene microspheres and thus the seeding density decreases with time. It has been observed that an initial particle density of 0.05 ppp reduces to about 0.03 ppp over seven hours in the case of non-rotating convection. In the present experimental apparatus, the working fluid is seeded before placing the cylinder in situ and it is not possible to add further seeding particles after the beginning of the experiment. This, in conjuction with the PC memory resources, limits the duration of the experiment; consequently, the longest run lasted about four hours.

Fluorescence of the particles is exploited to reduce the green reflections from the copper base and increase the particle contrast. For this purpose, the camera lenses are equipped with HOYA YA3 orange filters. Indeed, after the particle sedimentation, both the top and the bottom plates are covered by a layer of bright particles and this affects considerably the tracking of particles passing near the plates. As a consequence, the accuracy of the measurements in proximity of the plates is reduced; in addition, it should be commented that the seeding density is not high enough to resolve the boundary layers on both the plates and the cylinder sidewall.

Optical calibration of the camera system is a critical point of the present experimental setup, because of inaccessibility of the cylinder interior and optical distortions caused by the curvature of the sidewall. The beneficial refractive index-matching effect provided by the surrounding water in the tank was found to be insufficient to reach high accuracy in the tomographic reconstruction of the particle distribution via the classical pinhole camera model [149, 150]. To obviate this, a modification of the pinhole camera model has been introduced and experimentally validated. This innovative model is presented and diffusely discussed in chapter 4, where a comparison with other camera models is also carried out.

### 3.1.3. Rotating apparatus

FIGURE 3.5 reports a photo of the experimental rotating apparatus. A structure composed by Bosch aluminum profiles is mounted on a rotating table. The structure holds both the tank with the convection cell (A), the equipment (water circuits, TECs and power supplies) for the control of the top, bottom and tank bulk temperatures (B) and the four cameras (C). Also the computer controlling the image acquisition (G) is placed on the rotating table, while the laser (D-E) and the optics used to shape the cylindrical beam are on the ground. This is possible because the cylinder axis coincides with the rotation axis and, thus, after the laser beam is

Figure 3.5: Experimental rotating apparatus. A: water tank with the cylindrical sample; B: supplementary devices for temperature control (see figure 3.2); C: cameras; D: laser chiller; E: laser; F: optics for beam shaping; G: on-board PC for data acquisition; H: translational stage used in the camera calibration phase; I: rotating table.

centered on the cylinder, its position and orientation relative to the measurement volume does not change during the rotation.

A mercury slip ring with six conductors (Mercotac 630, 4-30 A per conductor, 0-250 V AC/DC) is installed on the rotating shaft for power transmission to the computer. Since the synchronizer that controls the timing of the cameras and the laser is also on the rotating table and is driven by the same computer, two of the channels of the slip ring are used to transmit the trigger signals to the laser.

The rotating table is driven by a synchronous brushless motor (STIMA MIRA 71b6 B5) with 6 poles (1000 rpm), a nominal power of $0.73\,\mathrm{kW}$ and an efficiency of $83.1\%$. The motor is piloted by an inverter (Sinus H 0002 4T BA2K2) that performs a vector control to ensure stability of the angular velocity. The rotation is transmitted through a gear box connected to the motor and a system of two wheels (Gates 8M-50S-21 and 8M-40S-21) with a toothed belt (the tooth depth is $5.9\,\mathrm{mm}$, while the tooth pitch is $8\,\mathrm{mm}$). The total gear ratio is approximately 1:31. Finally, a ball bearing provides the rotation to a square aluminum base plate on which the entire structure of figure 3.5 is screwed.

The rotation rate of the convection cell is varied between 0.5 and 1.3 rad/s in the present experiments. The rotation stability is measured by means of the gyroscopes integrated in an inertial measurement unit (SBG Systems IG-500N, GPS enhanced attitude and heading reference system) and the standard deviation of the rotation rate is found to be 0.5% of the rotation rate itself in the case of the slowest rotation investigated. Also tangential accelerations are found to be very small; the angular acceleration magnitude is below $10^{-2}$ rad/s$^2$ in all the tested conditions.

Before mounting the aluminum structure, the rotating table was leveled with a water-digital level; subsequently, also the convection cell is leveled to minimize angles between the cylinder axis and the rotation axis.

### 3.1.4. Image processing

IN all the investigated cases, measurements are carried out over about four hours with a sampling frequency of 7.5 Hz (corresponding to a sequence of 108,000 snapshots). This is sufficient to carry out a time-resolved analysis of the particle images and exploit the advantages of the motion enhanced approaches presented in the previous chapter. The image analysis consists essentially of three stages: image pre-processing, time-resolved motion analysis and velocity data post-processing.

Image pre-processing is aimed at eliminating undesirable image characteristics, such as constant and time-varying background intensity not associated to the particles and non-uniformity within an image and between cameras. This fundamentally affects the reconstruction quality and the accuracy of the velocity field. In the present case, non-uniformity of the light recorded by the different cameras is limited since the orientation of the cameras with respect to the cylindrical beam is the same and all the cameras record essentially the lateral scattering of the light from particles. Also time variations of the background intensity are considerably small since the two lasers are shot in the same camera frame in such a way to increase the recorded light intensity of the cylindrical beam. Non-uniformity within the image is affected by the reflections of the particle deposit on the two plates; indeed, without an appropriate pre-processing, the images of the particles passing in front of the plates are practically indistinguishable from the brighter background (after long time from the beginning of the experiment). Several techniques for the background noise removal were tested and it was found that the POD-based one [173] leads to the better results in comparison with other classical techniques, such as pixel-wise historical minimum subtraction or sliding minimum subtraction. After pre-processing, the quality factor of the tomographic reconstruction [174] is $\geq 0.89$ throughout the duration of the test.

Time-resolved motion analysis is based on a combination of the most recent algorithms for particle motion tracking and consists essentially of two steps. Initially, a first set of snapshots (typically 5-10) is analyzed by the sequential-motion-tracking enhancement (SMTE) algorithm [164]. At this stage, the light intensity distribution is reconstructed with multiple iterations of the SMART [157] and CSMART [175, 176] algorithms, using a multi-resolution approach [163]; multi-pass volumetric cross-correlations are performed with an efficient algorithm using sparse matrices [172]. This process is based on window deformation as explained in [177–180] for planar PIV. In the second phase of the process, the STB method [167] is used for particle tracking. Particle triangulation is performed by iterative particle identification [165], while forward-time projection of particles is based on both extrapolation of known trajectories and cross-correlation. Therefore, the present methodology falls in the class of the hybrid approaches that combine some characteristics of the T-PIV and the 4D-PTV techniques. However, most of the images are in fact processed with the latter technique and this allows a considerable saving of time.

The output of the above process consists of the particle tracks. The velocity data processing is aimed at estimating the instantaneous velocity fields from the particle trajectories. This implies estimation of the particle velocities and interpolation of the latter onto a structured grid. The particle velocity at a certain time instant is the slope of the particle trajectory at the same instant and is generally calculated with discrete derivative schemes. In doing this, small errors in the position of the particle result in large errors in the velocity and therefore temporal filters are typically employed to increase the accuracy of the velocity measurements. In the present experiments, a quadratic polynomial fitting based on a kernel of 5 time positions of the particles was used to calculate velocities. The employed kernel, with the current sampling frequency, corresponds to a time interval of $0.66$ s, which is smaller than the Kolmogorov time scale $\tau_\nu$ for the non-rotating case investigated here. The latter is $\tau_\nu = \sqrt{\nu/\varepsilon_u}$ with $\varepsilon_u$ the kinetic energy dissipation that can be estimated via the exact relationship $\varepsilon_u = \nu^3 L^{-4}(Nu - 1)RaPr^{-2}$. In the operating conditions of the non-rotating experiment presented below, $Nu$ is calculated from the numerical simulations and, being the remaining parameters known, $\tau_\nu$ is estimated to be about $1.8$ s. Note however that the employed relationship holds for the case of a cylinder with adiabatic sidewall. Moreover, although we have sufficient temporal resolution to filter trajectories without smoothing high frequencies, the spatial resolution is not so high to resolve dissipation. In fact, using the same relationship above, the Kolmogorov length scale $\eta_\nu = (\nu^3/\varepsilon)^{1/4}$ is found to be $1.4$ mm. In the experiments, with a seeding density of about $0.04$ ppp, only $65,000$ particles (with trajectories longer than $5$ sampling times) are reconstructed in the measurement volume. Supposing that these particles are uniformly distributed over the cylinder interior, in a volume of $\eta_\nu^3$ this corresponds to a number of particles of $0.275$. However, the following analysis is focused mainly on the behavior of the coherent structures of the RB convection rather than that of the small-scale turbulent fluctuations; in this regard, the spatial resolution of the current measurements is enough to capture the main features of the flow dynamics.

As concerns the interpolation of the particle velocities onto a structured grid (i.e., transformation from a Lagrangian reference frame to an Eulerian one) different approaches are possible, among which are collocation methods, inverse distance weighting average and interpolation based on least-squares polynomial fitting of local data. In the current investigation, only the latter method is used. In the present implementation, a structured (Cartesian or cylindrical) grid is chosen within the measurement volume; for each point of the grid, particles falling within a fixed search radius are identified and the velocities of such particles are used to determine a local polynomial fitting function which is then evaluated at the location of the grid point. Moreover, to reduce the effects of the ghost particles on the determination of the velocity field, only particles with trajectories longer than a fixed number of time instants are used in the above procedure. For the present experiments, we use a search radius of 30 voxels (about 2 mm) and a second-order polynomial function for the fit and employ only particles with trajectories longer than 7 time instants.

### 3.1.5. Experimental parameter settings

THE current work presents the results of the T-PIV measurements of RB convection in one specific non-rotating configuration and two rotating configurations differing for the value of $Ro$. The experimental parameter settings for such cases are summarized in Table 3.1.

Table 3.1: Experimental parameter settings.

| | parameter | value (non-rotating; rotating) | unit |
|---|---|---:|---|
| Cylindrical sample | material | Plexiglas (PMMA) | |
| | height, $L$ | 148 | mm |
| | diameter, $D$ | 74 | mm |
| | sidewall thickness, $t_s$ | 3 | mm |
| | sidewall density, $\rho_s$ | 1190 | kg/m$^3$ |
| | sidewall specific heat, $C_{p_s}$ | 1470 | J/(kg K) |
| | sidewall thermal conductivity, $\kappa_s$ | 0.19 | W/(mK) |
| | bottom temperature, $T_b$ | 20; 26 | °C |
| | top temperature, $T_t$ | 15; 21 | °C |
| | tank temperature (sidewall external side), $T_e$ | 17.35; 23.5 | °C |
| Fluid | type | water, liquid | |
| | average temperature, $T_m$ | 17.5; 23.5 | °C |
| | density, $\rho_f$ | 998.8; 997.5 | kg/m$^3$ |
| | specific heat, $C_{p_f}$ | 4183.5; 4180 | J/(kg K) |
| | thermal conductivity, $\kappa_f$ | 0.593; 0.604 | W/(mK) |
| | kinematic viscosity, $\nu$ | $1.08 \times 10^{-6}$; $0.924 \times 10^{-6}$ | m$^2$/s |
| | thermal diffusivity, $\alpha$ | $1.42 \times 10^{-7}$; $1.45 \times 10^{-7}$ | m$^2$/s |
| | thermal expansion coefficient, $\beta$ | $1.79 \times 10^{-4}$; $2.42 \times 10^{-4}$ | °C$^{-1}$ |
| Seeding particles | type | orange fluorescent polyethylene | |
| | diameter, $d_p$ | 58 | $\mu$m |
| | density, $\rho_p$ | 1000 | kg/m$^3$ |
| | relaxation time, $\tau_p$ | $1.73 \times 10^{-4}$; $2.02 \times 10^{-4}$ | s |
| | Stokes number[*], $St_p$ | $4.22 \times 10^{-5}$; $5.73 \times 10^{-5}$ | |
| | Reynolds number[*], $Re_p$ | 1.93; 2.62 | |

| | | | |
|---|---|---:|---|
| | Archimedes number, $Ar_p$ | $1.12 \times 10^{-4}$; $3.12 \times 10^{-4}$ | |
| | buoyancy Archimedes number[†], $Ar'_p$ | $4.08 \times 10^{-5}$; $7.50 \times 10^{-5}$ | |
| Laser | type | Nd:YAG, dual cavity | |
| | maximum pulse power | 200 | mJ |
| Cameras | type | sCMOS, 5.5 megapixel | |
| | employed sensor area | $2480 \times 1280$ | pixel |
| | pixel size | $6.5 \times 6.5$ | $\mu$m |
| Imaging | lens focal length, $f$ | 28 | mm |
| | lens aperture, $f_\#$ | 22 | |
| | view angles | $-60, -20, 20, 60$ | ° |
| | digital resolution | 14; 15 | pixel/mm |
| | size of reconstructed volume | $148 \times 80 \times 80$; $2072 \times 1120 \times 1120$; $2200 \times 1200 \times 1120$ | mm; pixel |
| Seeding concentration | particle per pixel, $ppp$ | 0.04; 0.03 | pixel$^{-2}$ |
| | mean particle spacing | 24; 32 | voxel |
| Acquisition | sampling frequency | 7.5 | Hz |
| | duration | 4; $\approx 3500$; $\approx 4080$ | hours; $\tau_f = L/u_0$ |
| | # of acquired snapshots | 108,000 | |
| Analysis | size of interrogation volumes for interp. | 60; 90; $2.9 \times 10^{-2}$; $4.0 \times 10^{-2}$ | voxel; $L$ |
| | final grid resolution ($N_\varphi \times N_r \times N_x$) | $69 \times 69 \times 99$ | |
| Rotation | rate, $\Omega$ | 0; 30, 75 | °/s |
| | stability (s.t.d.) | 0.5 | %$\Omega$ |
| Dimensionless parameters | aspect ratio, $\Gamma$ | 0.5 | |
| | Rayleigh number, $Ra$ | $1.86 \times 10^8$; $2.86 \times 10^8$ | |
| | Prandtl number, $Pr$ | 7.6; 6.4 | |
| | Rossby number, $Ro$ | $\infty$; 0.25, 0.1 | |
| | Froude number, $Fr$ | $0$; $1.03 \times 10^{-3}$, $6.46 \times 10^{-3}$ | |

[*] based on free-fall velocity $u_0$.
[†] calculated from equation (2.6) with $\theta = 0.5$.

## 3.2. Numerical method and procedure

THE numerical study of RB convection is carried out by solving the three-dimensional Navier-Stokes equations within the Boussinesq approximation, i.e. equations (1.11)-(1.13) presented in section 1.2.2. In these simulations, the presence of a physical sidewall is included. This implies solving the heat equation also in the domain constituted by the solid wall. Thus, we solve the following non-dimensional equations [112]:

$$\nabla \cdot \mathbf{u} \;=\; 0 \tag{3.1}$$

$$\partial_t \mathbf{u} + \mathbf{u} \cdot \nabla \mathbf{u} \;=\; -\nabla p + \sqrt{\frac{Pr}{Ra}} \nabla^2 \mathbf{u} - \theta \hat{\mathbf{g}} - \frac{1}{Ro} \hat{\boldsymbol{\Omega}} \times \mathbf{u} \tag{3.2}$$

$$\partial_t \theta + \mathbf{u} \cdot \nabla \theta \;=\; \sqrt{\frac{1}{Pr\,Ra}} \frac{(\rho C_p)_f}{(\rho C_p)_s} \nabla \cdot \left( \frac{\kappa_s}{\kappa_f} \nabla \theta \right) \tag{3.3}$$

where $(\rho C_p)_f$ and $(\rho C_p)_s$ are the heat capacity of the fluid and the sidewall and $\kappa_f$ and $\kappa_s$ are the thermal conductivity of the fluid and the sidewall, respectively. The adopted scaling is the free-fall one ($L$ and $\Delta$ are the length and temperature scales, while the velocity scale is the free-fall velocity $u_0$) and the centrifugal forces due to buoyancy have been neglected ($Fr = 0$). Note that $p$ is a reduced pressure separated from its hydrostatic contribution, but containing the centripetal contributions.

Equations (3.1-3.3) are solved in cylindrical coordinates; all the details about the numerical procedure are available in [57, 181, 182]. Equations (3.1-3.2) are solved over the fluid volume, i.e. for $0 \le x \le 1$, $0 \le r \le \Gamma/2$ and $0 \le \varphi < 2\pi$, while equation (3.3) is solved the overall domain, consisting of both the fluid and the sidewall, i.e. for $0 \le x \le 1$, $0 \le r \le \Gamma + t_s/L$ and $0 \le \varphi < 2\pi$, with $t_s$ being the thickness of the sidewall. No-slip condition is imposed at all the solid walls of the fluid domain, while constant temperature boundary condition is used at the bottom and top plates. Since the temperature field is solved on the whole domain no temperature boundary condition is required at the fluid/sidewall interface. Instead, a temperature boundary condition is imposed at the external side of the cylinder wall; in particular, here it is possible to assign both constant temperature and adiabatic condition. In all the simulations performed in the present thesis, we use the constant temperature condition. Temperature of the external side of the sidewall is considered uniform over the whole surface and its value is denoted by $T_e$ (or $\theta_e$ in non-dimensional terms).

In the simulations, it is supposed that the bottom and the top thermal sources extend below and above the sidewall. This geometry is consistent with the connection between the sidewall and the heat exchanger in the present setup, but not with that between the sidewall and the bottom copper insert (see figure 3.3b).

The employed computational grid is the same for all the simulations reported in this thesis and consists of $257 \times 97 \times 385$ nodes in the azimuthal, radial and axial directions, respectively. Nodes are clustered near the wall, where kinematic and thermal boundary layers are formed. The number of nodes was chosen based on the resolution criteria of Stevens *et al.* [183] for a fully resolved DNS and that of Shishkina *et al.* [184] for the minimal number of nodes that should be placed inside the boundary layers. 12 nodes are placed inside the sidewall.

For all the investigated cases, the flow is simulated for at least 300 dimensionless time units to make sure to have reached the statistically stationary state and all transient effects are washed out, then data is collected for at least 1,000 additional dimensionless time units (and in some cases even for $4{,}000$ dimensionless time units) so that the statistical convergence could be verified.

Calculations are carried out on SURFSara Cartesius supercomputer located in Amsterdam using $32$ cores for each run. The numerical parameter settings for the various simulations of the present work are reported in Table 3.2.

Table 3.2: Numerical parameter settings.

| | parameter | value (non-rotating; rotating) |
|---|---|---|
| Computational grid | azimuthal distribution type | uniform |
| | radial distribution type | Chebyshev |
| | axial distribution type | Chebyshev |
| | number of nodes $(N_\varphi \times N_r \times N_x)$ | $257 \times 97 \times 385$ |
| | time step | variable, max.$= 0.01$ |
| Statistical information | run duration | 1,300-4,300 time units |
| | averaging time | 1,000-4,000 time units |
| Boundary conditions | velocity on the plates and sidewall external side | no-slip |
| | temperature at the bottom plate, $\theta_b$ | 0 (uniform) |
| | temperature at the top plate, $\theta_t$ | 1 (uniform) |
| | temperature at the external side of the sidewall, $\theta_e$ | 0.48; 0.5 (uniform) |
| Initial condition | velocity | fluid at rest |
| | temperature | pure conductive steady distribution + sine perturbation |
| Dimensionless parameters | aspect ratio, $\Gamma$ | 0.5 |
| | Rayleigh number, $Ra$ | $1.86 \times 10^8$ |
| | Prandtl number, $Pr$ | 7.6 |
| | Rossby number, $Ro$ | $\infty$; 1, 0.5, 0.25, 0.15, 0.1, 0.075, 0.05 |
| | Froude number, $Fr$ | 0 |
| Sidewall properties | thickness, $t_s/L$ | 0.020270 |
| | density, $\rho_s/\rho_f$ | 1.1915 |
| | specific heat, $C_{p_s}/C_{p_f}$ | 0.41866 |
| | thermal conductivity, $\kappa_s/\kappa_f$ | 0.32031 |

# 4

# Camera calibration model for imaging through a cylinder

## 4.1. Introduction

THIS chapter presents a novel camera calibration model designed to account for the optical deformation of the image caused by the cylinder sidewall.

Such a distortion is related to a twofold refraction of the optical rays passing through the cylindrical wall, as shown in figures 4.1a: a first refraction occurring at the external surface of the wall and a second one occurring at its internal surface. Both the refractions follow Snell's law (figure 4.1b):

$$n_i \sin \vartheta_i = n_r \sin \vartheta_r \tag{4.1}$$

where $n_i$ and $n_r$ are the refractive indexes of the medium transmitting the incident ray and the medium transmitting the refracted ray, respectively, while $\vartheta_i$ and $\vartheta_r$ are the angles between the normal of the interface and the incident and refracted rays. Equation (4.1) shows that the deviation of the optical ray depends essentially on the ratio $\varrho = n_i/n_r$ between the refractive indexes of the two optical media and the incident angle $\vartheta_i$. In particular, the incident ray normal to the external wall of the cylinder ($\vartheta_i = 0$) does not undergo deviation, while the remaining rays are diverted from their original directions.

It is worth noting that although Snell's law is typically formulated for light rays, it applies to LOSs as well by virtue of the principle of optical reciprocity. In other words, it is possible to imagine that the optical rays drawn in figures 4.1a are either the light rays emerging from a point source or the LOSs starting from the optical center of an imaging system placed outside the cylinder. For the discussion to follow, we are interested in the latter case.

The situation depicted in figures 4.1a is similar to the case of a Plexiglas wall submerged in water (case under investigation), that is to say the optical media outside and inside the cylinder are identical and their refractive index is lower than that of the cylindrical wall. This results in an opposite deviation of the LOSs at the external and the internal side of the sidewall; more specifically, the deviation at the

53

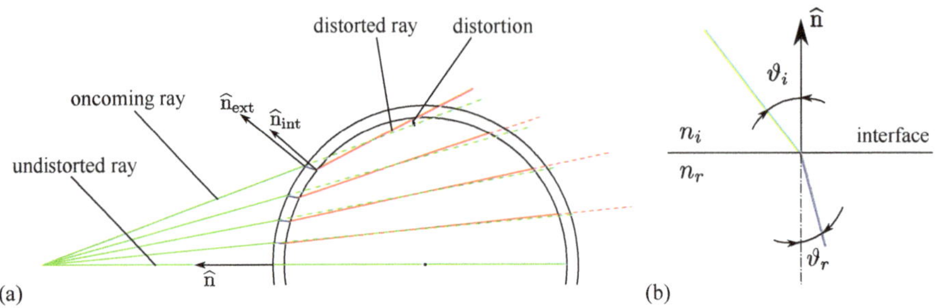


Figure 4.1: Schematic representation of the distortion of the optical rays through the cylinder sidewall: (a) view in the direction of the cylinder axis; (b) local refraction of the incident ray at the interface between the fluid and the wall.

external side results in a decrease of the angle between the LOS and the surface normal ($\vartheta_r < \vartheta_i$), whereas the deviation at the internal side leads to an increase of this angle ($\vartheta_r > \vartheta_i$). It is then evident that the distortion evaluated as distance of a point on the distorted ray from the extension of the incident ray is strongly variable with the position inside the cylinder and generally larger in the regions adjacent to the wall.

The above-discussed problem is relevant to a broad range of optical investigations on fluid flows inside cylinders, among which, apart from RB convection, are the wall-bounded turbulence in pipes [185, 186], the in-cylinder flows of internal combustion and diesel engines [187, 188] and the Taylor-Couette flow [189, 190]. In these cases, the most common solution adopted to reduce the effect of the cylindrical deformation is based on refractive index matching techniques. These imply an appropriate design of the experimental facility since the cylinder has to be inserted in a box filled with the index-matching fluid (see for instance [95, 191]) or, alternatively, optical prisms filled with this fluid have to be attached directly to the cylinder sidewall (as in [116]). However, in most situations, the optical deformation is not completely removed since the internal fluid is required to have specific physical properties and it is not possible to achieve a perfect matching between its refractive index and that of the cylindrical wall. In such a way, only the deviation of the LOSs at the external side of the cylindrical wall is obviated, while that occurring at the internal side remains. Obviously, the larger the curvature (i.e., the smaller the internal radius), the greater the latter deviation.

In T-PIV applications, the deformation caused by the cylinder sidewall plays a fundamental role in the tomographic reconstruction step. As a matter of fact, the camera calibration model has to account for it to accurately describe the correspondence between the 3D world coordinates and the 2D image ones. As mentioned in section 2.4, the most common camera models are the Tsai-Heikkila pinhole camera model and the polynomial functions. The former has the advantage of being very simple and consisting of a small number of calibration parameters, but on the downside, it does not include the effects of distortions along the LOSs (some corrections are indeed introduced in the model to account for distortions in the direction normal to the LOSs). The cylindrical deformation is among these kinds of distortion and, therefore, the pinhole camera model cannot be applied to precisely map the inner volume of a transparent cylinder without any modification. Conversely, polynomial mapping functions work in principle as a Taylor series expansion and can describe any sufficiently smooth correspondence between the 3D world and the 2D image coordinates, accounting for effects like optical misalignment, thermal deformation and refraction by optical windows or fluid interfaces. The main downside of the polynomial models is that they require a very high number of constants when large volumes are mapped or moderate to strong distortions are present in the field of view. This results in very high-degree polynomials that can introduce local oscillations leading to high reconstruction errors. Alternatively, piecewise polynomial interpolating functions (as the multi-plane polynomial model, see section 2.4) may be used, but this increases the complexity of the model and the computational burden related to the operations of inverse mapping (from 2D images to 3D world).

In this work, an innovative camera calibration model is introduced to overcome the limits of both the pinhole camera model and the polynomial models. Such a model relies on the pinhole camera model and preserves its simplicity by integration of Snell's laws to model the refraction of the optical rays at the external and internal surfaces of the cylindrical wall. Making use of the perspective and refraction laws, the mapping function consists of a relatively small number of parameters and all of these parameters have a clear geometrical or physical meaning. This model offers the additional advantage of allowing a calibration procedure that does not require the sweeping of a target in the cylinder interior. This feature is of paramount importance because of the inaccessibility of the inner volume of the convection cell in the present experimental apparatus.

In the following, firstly the pinhole camera model is briefly introduced and its limits in mapping the inner volume of the cylindrical sample are discussed, secondly, the novel model is presented, then an effective calibration procedure is outlined. Finally, the novel model is comparatively assessed against other polynomial models by using experimental data.

## 4.2. Pinhole camera model

### 4.2.1. Definition of the model

$T$HE pinhole camera model is based on a pure perspective projection model. An ideal pinhole camera is characterized by a point aperture that focuses the light rays from the field of view without the aid of lenses; these rays form an inverted image of the scene on a photographic film or a digital sensor placed behind the aperture (which is the center of projection). Pinhole cameras are not suitable for imaging in low light conditions or recording of fast moving objects due to their very small aperture. Therefore, digital cameras typically feature sufficiently large aperture and use spherical lenses to focus light without blurring. Similarly to a pinhole, the imaging of a high quality lens can be still described by a perspective projection model, as that shown in figure 4.2 (here, the lens is not represented for clearness). However, the imperfections of the lens introduce some distortions (greater at the edges of the

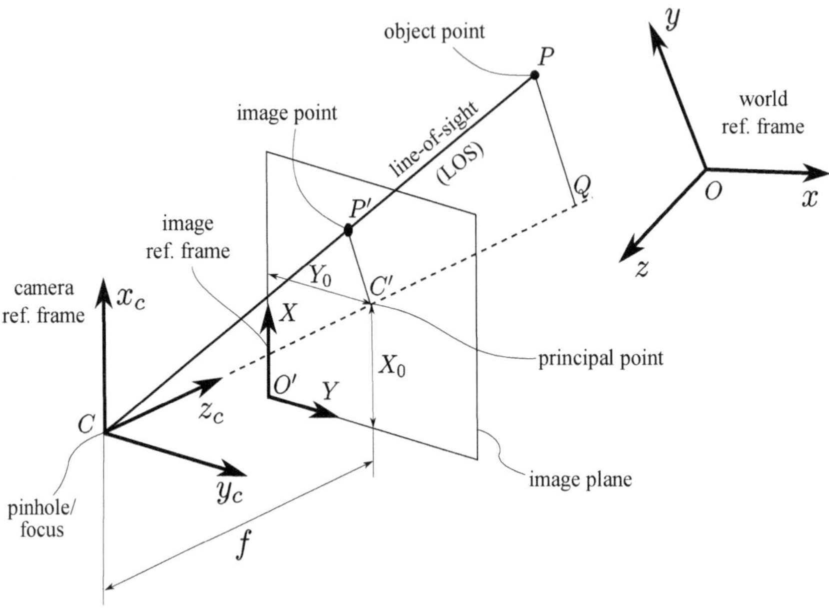

Figure 4.2: Pure perspective projection model.

lens itself) that have to be appropriately included in the camera model.

Referring to figure 4.2, it is possible to identify three different reference frames: the world reference frame $Oxyz$, the image reference frame $O'XY$ and the camera reference frame $Cx_cy_cz_c$. $C$ is the center of the projection (pinhole or lens focus) and the $x_c$- and $y_c$-axes are chosen to be parallel to the $X$- and $Y$-axes for simplicity, whereas the $z_c$-axis (*principal axis*) is normal to the image plane. The intersection $C'$ of the principal axis with the image plane (orthogonal projection of $C$ on the image plane) is named *principal point* and its coordinates in the image reference frame are $[X_0, Y_0]^T$. The distance $CC'$ is the focal length and is denoted with $f$. The LOS of the object point $P$ is the line $CP$ and its intersection with the image plane is the image point $P'$. The orthogonal projection of $P$ on the principal axis is indicated with $Q$.

The perspective transformation is simply given by:

$$\frac{P' - C'}{f} = \frac{P - Q}{|Q - C|} \tag{4.2}$$

the projections of which along the $x_c$- and $y_c$-axes yield:

$$\begin{bmatrix} X \\ Y \end{bmatrix} = \frac{f}{z_c} \begin{bmatrix} x_c \\ y_c \end{bmatrix} + \begin{bmatrix} X_0 \\ Y_0 \end{bmatrix} \tag{4.3}$$

where $[X, Y]^T$ are the image coordinates of $P'$ (in physical units) and $[x_c, y_c, z_c]^T$

are the coordinates of $P$ in the camera reference frame. The latter are related to the world coordinates of $P$ by the geometric transformation:

$$\begin{bmatrix} x_c \\ y_c \\ z_c \\ 1 \end{bmatrix} = \begin{bmatrix} \mathbf{R} & \mathbf{t} \\ 0 & 1 \end{bmatrix} \begin{bmatrix} x \\ y \\ z \\ 1 \end{bmatrix} = \mathbf{T} \begin{bmatrix} x \\ y \\ z \\ 1 \end{bmatrix} \tag{4.4}$$

with $\mathbf{R}$ being a rotation matrix defined by three Euler angles $\phi$, $\delta$ and $\psi$ and $\mathbf{t} = [t_x, t_y, t_z]^{\mathrm{T}}$ the translation of the origins between the camera frame and the world frame.

Ultimately, in homogeneous coordinates the projection transformation from the 3D world coordinates to the 2D image ones can be written as:

$$c \begin{bmatrix} X \\ Y \\ 1 \end{bmatrix} = \mathbf{KT} \begin{bmatrix} x \\ y \\ z \\ 1 \end{bmatrix} \tag{4.5}$$

where $c$ is a scaling constant and $\mathbf{K}$ is the perspective projection matrix given by:

$$\mathbf{K} = \begin{bmatrix} \chi f / d_{\mathrm{pix}} & 0 & X_0 & 0 \\ 0 & f / d_{\mathrm{pix}} & Y_0 & 0 \\ 0 & 0 & 1 & 0 \end{bmatrix}. \tag{4.6}$$

In equations (4.5-4.6), the image coordinates are now expressed in pixel units, $d_{\mathrm{pix}}$ is the pixel size in the $Y$-direction and $\chi$ is the ratio of this size to the pixel size along the $X$-direction and has been introduced to account for the image stretching due to a rectangular shape of the sensor pixels. The parameters $t_x$, $t_y$, $t_z$, $\phi$, $\delta$, $\psi$ are called the extrinsic parameters of the pinhole camera model, while $f$, $X_0$, $Y_0$, $\chi$, $d_{\mathrm{pix}}$ are the so-called intrinsic parameters.

As previously mentioned, the pinhole camera model includes some corrections to account for unavoidable distortions affecting the imaging system. A first source of distortion is the non-orthogonality of the camera sensor that is corrected by multiplying the matrix $\mathbf{K}$ by the matrix:

$$\mathbf{B} = \begin{bmatrix} 1 + b_1 & b_2 & 0 \\ b_2 & 1 - b_1 & 0 \\ 0 & 0 & 1 \end{bmatrix}. \tag{4.7}$$

where $b_1$ and $b_2$ are the so-called linear distortion coefficients. Such a correction is explicit in the sense that it does not depend on the position of the image points and is simply obtained by linearly combining the undistorted image coordinates. Secondly, distortions are caused by the imperfections of the lenses used to focus the light. Following Weng *et al.* [192] it is possible to identify three main types of geometrical lens distortions, namely the radial distortions, the decentering distortions and the thin prism distortions. Each of this type of distortion depends on the position of the image point and the resulting total distortion is given by:

$$\delta X(X,Y) \;=\; X(k_1 r^2 + k_2 r^4) + p_1(r^2 + 2X^2) + 2p_2 XY + s_1 r^2 \qquad (4.8)$$

$$\delta Y(X,Y) \;=\; Y(k_1 r^2 + k_2 r^4) + 2p_1 XY + p_2(r^2 + 2Y^2) + s_2 r^2 \qquad (4.9)$$

where $k_1$, $k_2$ are the coefficients for the radial distortions, $p_1$, $p_2$ those for the decentering distortions and $s_1$, $s_2$ those for the thin-prism deformation. Note that terms of order greater than the second have been retained only for radial distortions (terms weighted by the coefficient $k_2$). Moreover, the thin prism distortion is commonly neglected [150]. Therefore, the inclusion of a distortion model adds six intrinsic parameters ($b_1$, $b_2$, $k_1$, $k_2$, $p_1$, $p_2$), for a total of 17 parameters.

It is worth noting that the image position correction given by terms in equations (4.8-4.9) is implicit (i.e., dependent on the same image position) since the corrected correspondence between the 3D world coordinates and the 2D image coordinates is:

$$c \begin{bmatrix} X + \delta X(X,Y) \\ Y + \delta Y(X,Y) \\ 1 \end{bmatrix} = \mathbf{BKT} \begin{bmatrix} x \\ y \\ z \\ 1 \end{bmatrix}. \qquad (4.10)$$

From equation (4.10), starting from the 3D world coordinates $[x,\,y,\,z]^{\mathrm{T}}$ the image coordinates $[X,\,Y]^{\mathrm{T}}$ can be determined iteratively by non-linear least squares optimization algorithms. Since this procedure is computationally expensive and distortions are typically very small, in most situations an explicit image correction is indeed applied. Therefore, first the pair of ideal undistorted coordinates $[X_i,\,Y_i]^{\mathrm{T}}$ is determined from equation (4.5), then the correction for the image distortions is calculated by using such coordinates as $[\delta X(X_i,Y_i),\,\delta Y(X_i,Y_i)]^{\mathrm{T}}$ and finally the corrected, distorted coordinates are obtained as $[X_i - \delta X(X_i,Y_i),\,Y_i - \delta Y(X_i,Y_i)]^{\mathrm{T}}$. The pinhole camera model used in the present work relies always on explicit image correction for the lens distortions.

## 4.2.2. Limits of the pinhole approximation

THE two basic assumptions of the pinhole camera model are that the LOSs are straight lines and these lines converge in a unique point, the center of projection. When one of these two conditions does not hold, the model is no longer suitable for describing the correspondence between the 3D world and the 2D image coordinates. In the present experimental apparatus, there are basically three factors that can compromise the applicability of the pinhole camera model:

1. the local variations of the refractive index of the convective fluid with temperature, that might cause curvature of the LOSs;

2. the distortion of the LOSs through the Plexiglas windows of the tank, i.e. the refraction experienced by the optical rays passing from air to Plexiglas and from Plexiglas to water;

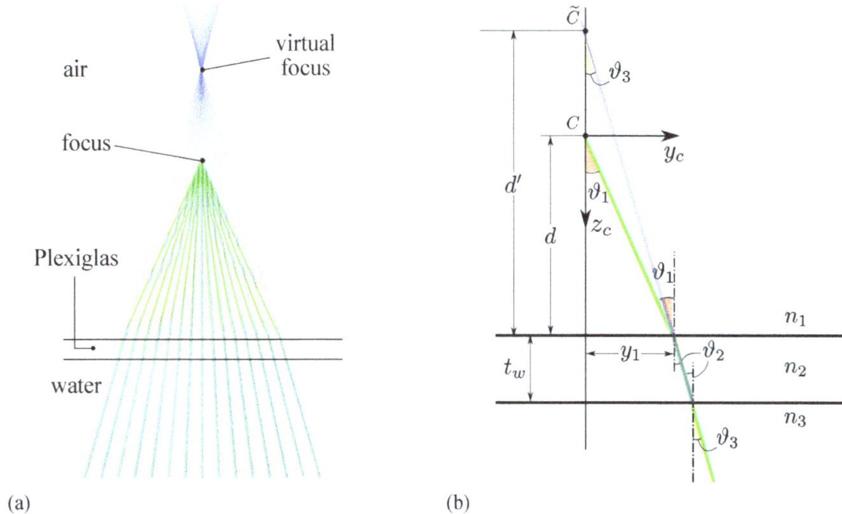


Figure 4.3: Deviation of optical rays through a plane wall of finite thickness: (a) example corresponding to the case of a Plexiglas wall separating air and water (the focal point is in air); (b) sketch of the geometry of refraction at both interfaces.

3. the optical deformation caused by the cylinder sidewall, as diffusely discussed above.

In order to asses the importance of the first effect, we use a procedure similar to that of Ni *et al.* [116]: a calibration target is placed behind the cylinder (it is not possible to place it inside the cell) and the displacement of the control points over the time is evaluated. The rms value of such displacements in the image plane is estimated to be lower than 0.01 pixel (which is significantly below the search radius used in the triangulation for IPR) and this confirms that the effects of the temperature gradients can be neglected from the viewpoint of the identification of the cameras' mapping functions. Furthermore, since a temporal filter (see section 3.1.4) is used to estimate the particle positions, velocities and accelerations, these effects are expected to affect to no extent the accuracy in the estimation of the dynamic quantities of the flow, as shown by the analysis of Ni *et al.* [116].

As concerns the second effect, figure 4.3a shows the distortion of the optical rays passing through a plane wall of finite thickness made of Plexiglas and separating air from water. The deviation of the rays has been calculated numerically by applying Snell's law and assuming refractive index of air, Plexiglas and water respectively equal to 1, 1.49 and 1.33. It is possible to note that the extensions of the refracted rays (blue transparent lines) almost converge in a virtual focus, placed at a distance from the wall greater than that of the actual focus. This property can be readily demonstrated analytically. For this purpose, consider figure 4.3b. Here, the actual focus is denoted with $C$ and its distance from the interface between the media 1 and 2 is indicated with $d$; $\widetilde{C}$ represents the intersection of the extension of the twice-refracted ray (i.e., the ray propagating in the medium 3) with the $z_c$-axis. Each ray can be identified by the distance of its intersection point with the interface 1-2 from the $z_c$-axis, which is denoted with $y_1$. To prove that all the extensions of the refracted rays converge in the point $\widetilde{C}$ it is sufficient to demonstrate that $d'$ is independent of $y_1$. If it is supposed that $d \gg y_1$, than also $d' \gg y_1$ and the following relationships hold:

$$d \;=\; \frac{y_1}{\tan \vartheta_1} \approx \frac{y_1}{\vartheta_1} \tag{4.11}$$

$$d' \;=\; \frac{y_1}{\tan \vartheta_3} \approx \frac{y_1}{\vartheta_3}. \tag{4.12}$$

From the above equations it follows that:

$$d' = d\,\frac{\vartheta_1}{\vartheta_3} \tag{4.13}$$

while the application of Snell's law to the refraction at the interfaces 1-2 and 2-3, by virtue of the smallness of the angles $\vartheta_1$, $\vartheta_2$ and $\vartheta_3$, gives:

$$n_1\,\vartheta_1 = n_2\,\vartheta_2 = n_3\,\vartheta_3. \tag{4.14}$$

Combining equations (4.13-4.14) finally yields:

$$d' = d\,\frac{n_3}{n_1} = d\,\varrho_{1-3} \tag{4.15}$$

and this proves that $d'$ does depend only on the distance $d$ of the actual focus from the external side of the plane wall and the ratio $\varrho_{1-3}$ of the refractive indexes of the media separated by the wall. Interestingly, the location of the virtual focus depends neither on the material constituting the wall (i.e., the refractive index $n_2$) nor the wall thickness $t_w$.

The existence of a virtual focus for the refracted rays means that the pinhole-camera model can be employed to map the region 3 of figure 4.3b, although by adopting as center of projection the virtual focus $\widetilde{C}$ instead of the actual focus $C$. Nevertheless, such an approximation holds provided that the angles of the optical rays to the window normal are sufficiently small. In fact, the refracted rays corresponding to incident rays characterized by large $\vartheta_1$ do not converge in the virtual focus $\widetilde{C}$. However, the deviation of these rays from $\widetilde{C}$ remains sufficiently small even when the approximation $\vartheta_1 \approx \tan \vartheta_1 \approx \sin \vartheta_1$ does not hold and this deviation is partially compensated by the radial distortion model. In the present experiments, the camera sensor is nearly parallel to the Plexiglas window between air and water and only a small portion of the latter is used to image the field behind; despite the short focal length of the camera lenses, the maximum value of the angle between the LOS

   |   CHAPTER 4. Camera calibration model for imaging through a cylinder

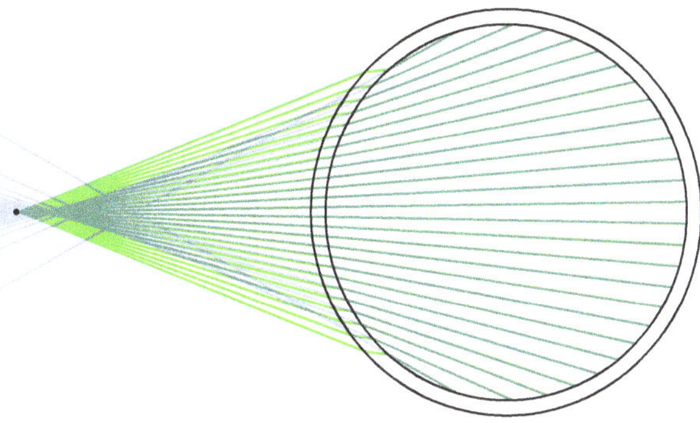

Figure 4.4: Deviation of optical rays through a cylindrical wall of finite thickness (case of a Plexiglas wall submerged in water).

and the window normal is estimated to be $6°$ which is small enough for the above approximation to hold.

The inadequacy of the pinhole camera model in mapping the cylinder interior is evident in figure 4.4. In fact, differently from the case of the plane wall, the extensions of the twice-refracted rays do not merge in a virtual focus. In particular, the rays intersecting the lateral parts of the cylinder undergo stronger deviations and diverge significantly from the remaining rays. This also elucidates that the cylindrical distortion is more severe in the regions adjacent to the cylinder wall. In conclusion, a modification of the pinhole camera model is needed to map accurately the cylinder interior.

## 4.3. Pinhole camera model with refraction correction for imaging through a cylinder

THE optical distortion caused by the cylinder sidewall can be included in the pinhole camera model by using Snell's law to model the refractions of the LOSs at both sides of the sidewall itself. The corresponding imaging model is represented in figure 4.5. The main difference from the model in figure 4.2 is the addition of an optical cylindrical element of internal radius $r_i$ and thickness $\Delta r$. $G\xi\eta\zeta$ is a reference frame attached to the cylinder with the $\xi$-axis coincident with the cylinder axis. Due to the axisymmetry, the $\eta$- and $\zeta$- axes can be chosen arbitrarily. Moreover, for the discussion to follow, it is assumed that the cylinder extends infinitely along the $\xi$-direction. If the following treatment can be applied in principle also to the case of a finite-height cylinder, the assumption of an infinite cylinder, which is always acceptable when interested in mapping only the cylinder interior, greatly reduces the computational burden. In the case of an infinite cylinder, also the location of $G$ can be chosen in an arbitrary way, provided that it lies on the cylinder axis.

With the above assumptions, the location and orientation of the cylinder with

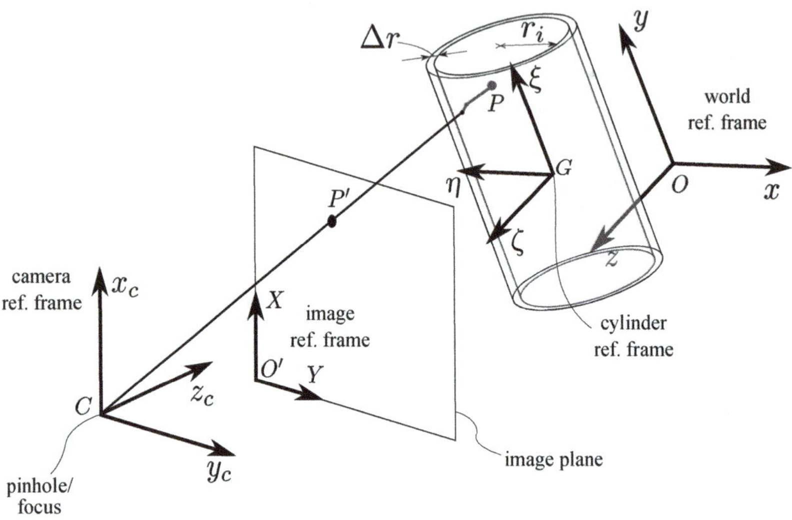


Figure 4.5: Perspective projection model with inclusion of a cylindrical optical element.

respect to the world reference frame are identified by only four parameters, namely two translations and two Euler angles. This can be readily understood by considering the transformation from the cylinder reference frame to the world reference frame shown in figure 4.6. Such a transformation is obtained by three consecutive rotations $\gamma_c$, $\beta_c$ and $\alpha_c$ about the $\xi$-, $\eta$- and $\zeta$-axes respectively and the origin translations $\xi_0$, $\eta_0$ and $\zeta_0$ about the same directions. Due to the axisymmetry, $\gamma_c$ can be chosen arbitrarily, as well as $\xi_0$ can be assigned any value by virtue of the infinite height of the cylinder. Obviously, the simplest choices are $\gamma_c = 0$ and $\xi_0 = 0$. In such a way, the coordinate transformation between the two reference frames is:

$$\begin{bmatrix} \xi \\ \eta \\ \zeta \end{bmatrix} = \begin{bmatrix} \cos(\alpha_c) & \sin(\alpha_c) & 0 \\ -\sin(\alpha_c) & \cos(\alpha_c) & 0 \\ 0 & 0 & 1 \end{bmatrix} \begin{bmatrix} \cos(\beta_c) & 0 & \sin(\beta_c) \\ 0 & 1 & 0 \\ -\sin(\beta_c) & 0 & \cos(\beta_c) \end{bmatrix} \begin{bmatrix} x \\ y \\ z \end{bmatrix} + \begin{bmatrix} 0 \\ \eta_0 \\ \zeta_0 \end{bmatrix} \qquad (4.16)$$

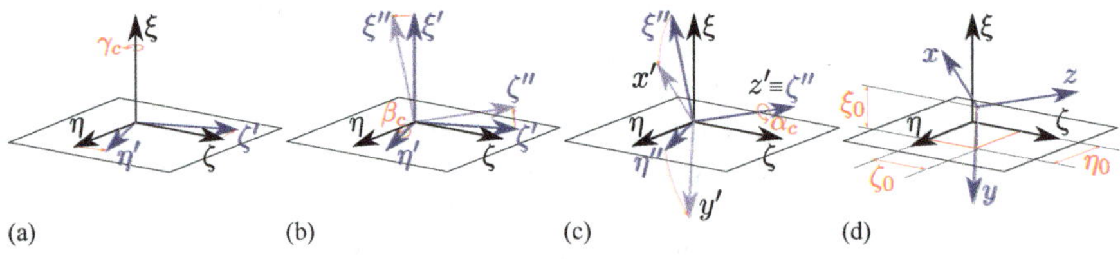


Figure 4.6: Geometric transformation from the cylinder reference frame $G\xi\eta\zeta$ to the world reference frame $Oxyz$: rotations about (a) the $\xi$-axis, (b) the $\eta$-axis and (c) the $\zeta$-axis and (d) origin translation.

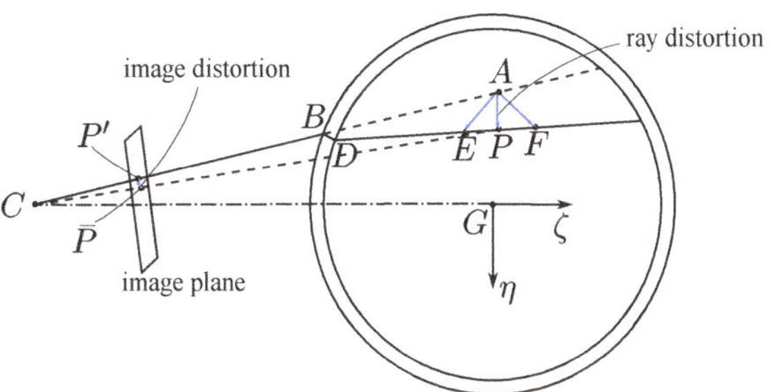


Figure 4.7: Definition of the cylindrical deformation. Distortion is exaggerated for clarity.

from which it is evident that the relative position and orientation of the cylinder depends only on four parameters ($\alpha_c$, $\beta_c$, $\eta_0$, $\zeta_0$). These parameters are called the *cylinder extrinsic parameters*.

As concerns the cylinder geometry, it is identified uniquely by the internal radius $r_i$ and the thickness $\Delta r$ (when assuming an infinite height). These parameters, along with the extrinsic parameters, contribute to identify the three-dimensional position of the internal and external surfaces of the cylindrical element, which represent the interfaces external fluid/sidewall and sidewall/internal fluid. The refraction of the LOSs occurring at each of these interfaces depends, as shown by equation (4.1), also on the ratio of the refractive indexes of the two confining material media. We indicate this ratio with $\varrho_e$ for the external surface and $\varrho_i$ for the internal surface. In the present experiments, $\varrho_e = 1/\varrho_i$. The parameters $r_i$, $\Delta r$, $\varrho_e$ and $\varrho_i$ are the so-called *cylinder intrinsic parameters*.

Let us now give an unambiguous definition of the cylindrical distortion. For this purpose, we refer to figure 4.7, where it has been supposed that the $\zeta$-axis is coincident with the line connecting the projection center $C$ and the origin $G$ of the cylinder reference frame. Note that in figure 4.7 the distortion is represented in a view parallel to the cylinder axis and, since in general the image plane is not parallel to the cylinder axis, its projection onto this view is a parallelogram. The figure shows that (neglecting lens distortions) the point $P$ is imaged along the LOS $PEDBC$ and its image is the point $P'$. Without the cylinder, $P$ would be imaged along the line $PC$ and its image would be the point $\bar{P}$, while the LOS corresponding to the point $P'$ would be the line $AC$. The image distortion caused by the cylinder is given by the vector $P' - \bar{P}$. Once determined such a distortion, the image point $\bar{P}$ in absence of the cylinder can be computed by applying the classical pinhole camera model (see 4.2) and then the actual image point $P'$ is obtained simply by adding the distortion itself.

It is now noted that the determination of the image distortion $P' - \bar{P}$ is possible only in an implicit way. In fact, given the image point $P'$, the LOS $AC$ is identified and it is immediate to determine the distorted LOS $PEDBC$ by using Snell's law. This implies:

1. determining the intersection $B$ between the line $AC$ and the external surface of the cylinder;

2. applying Snell's law to determine the deviation of the LOS $BC$ through the interface external ambient/sidewall;

3. determining the intersection $D$ between the distorted LOS propagating through the sidewall and the internal surface of the cylinder;

4. applying Snell's law to determine the new deviation of the LOS $DB$ through the interface sidewall/cylinder interior.

Conversely, given the object point $P$, the LOS $PEDBC$ is not immediately identified. In this case, the problem is finding among the infinite rays that start from $P$ the one that reaches the projection center $C$. Obviously, by virtue of the principle of the optical reciprocity, this is equivalent to find among the infinite rays that start from $C$ the one that reaches the object point $P$. We approach the problem from this second point of view.

By focusing on the LOS $AC$ in figure 4.7, we can define a ray distortion as the distance of the point $A$ from the distorted LOS $PEDBC$. If the easiest way to measure this distance is in the plane normal to one of the two LOSs, in principle it is possible to measure it in any plane except that parallel to the distorted LOS. For instance, the vectors $E - A$, $P - A$ and $F - A$ are all representative of the LOS deviation and can be assumed as ray distortion indifferently. In the present case, the ray distortion is evaluated in the plane passing through $A$ and normal to the $\zeta$-axis; therefore, it is coincident with the vector $P - A$. We denote such a distortion with the symbol $\delta_c P$.

Therefore, given the point $A$, it is possible to determine the distorted LOS $PEDBC$ similarly to the case when the image point $P'$ is assigned, as above explained, calculate the ray distortion $\delta_c P$ and then the "distorted" position $P$ as:

$$P = A + \delta_c P(A). \tag{4.17}$$

Vice versa, given the object point $P$, an iterative procedure has to be applied to determine the position $A$ from equation (4.17). In particular, we use a Newton-Raphson method starting from the initial guess $A = P$, which corresponds to the case of zero distortion. Once determined $A$, the image point $P'$ is readily obtained by applying the classical pinhole camera model.

In short, the above described model is based on straightening the LOSs of the points imaged through the cylinder. This is done by subtracting the cylindrical ray distortion from the position of the object points and subsequently applying the perspective projection transformation. The determination of the ray distortion is performed in the cylinder reference frame: in such a way it is possible to reduce the computational burden since both the intersections line/cylindrical surface and the application of Snell's law are cheaper when the cylinder axis coincides with one of the coordinate directions. To sum up, the application of the model consists of the following steps:

1. the coordinates of the object point $P$ are transformed from the world reference frame to the cylinder reference frame via the direct transformation (4.16);

    CHAPTER 4. Camera calibration model for imaging through a cylinder

2. the point $A$ on the undistorted LOS is determined by solving equation (4.17) iteratively via a Newton-Raphson method with initial guess $A = P$;

3. the coordinates of $A$ are transformed back to the world reference frame via the inverse transformation (4.16);

4. the image point $P'$ is obtained by applying the perspective projection transformation with corrections for distortion (4.10) of the classical pinhole camera model to the point $A$.

Although the above treatment has focused on the projection of points inside the cylinder, the present model can be applied to map the whole region surrounding the cylinder sidewall. In fact, the points that are not imaged through the cylinder are characterized by a zero ray distortion ($\delta_c P(A) = 0$) and therefore consistent with the model. For the points behind the cylinder, the ray distortion can be computed by determining the deviation of the LOSs through the sidewall both when entering and when exiting the cylinder. This implies a distinction between different possible cases, since some LOSs can intersect the external side of the cylinder but not the internal one and exit from the cylinder propagating entirely through the sidewall. In principle, also the points falling inside the sidewall can be mapped by the model. Nevertheless, it should be noted that some LOSs intersecting the external surface undergo total internal reflection and following their path may become very complicate. In such a case, the ray distortion is assigned a value equal to zero; this ensures that in the step 2 of the application of the model, the iterative procedure converge to the solution at the first iteration. Although this results in the projection of a point that is not actually visible, it does not represent a problem since the interior of the sidewall is generally not imaged. On the other side, imaging of the points behind the cylinder is used in the calibration of the camera system as explained in the next section.

## 4.4. Calibration procedure

THE camera calibration based on the pinhole camera model with the refraction correction for imaging through a cylinder consists essentially of two steps: identification of the pinhole-camera parameters and identification of the cylinder parameters. If possible, it is convenient to perform these two steps separately in order to facilitate the convergence of the optimization algorithms used for the estimation of the same parameters.

The pinhole-camera parameters can be determined via a classical target-based calibration, as shown in Figure 4.8a. The target is swept through the measurement volume along its normal direction in absence of the cylinder and its images at different depths are recorded by all the cameras. From the analysis of such images, it is possible to determine the pinhole-camera parameters for each camera with a procedure similar to [150]. Such a procedure consists of two steps: first, the parameters of the perspective projection transformation and the linear distortion $b_1$ and $b_2$ are estimated using a direct linear transformation (DLT), then iterative methods are

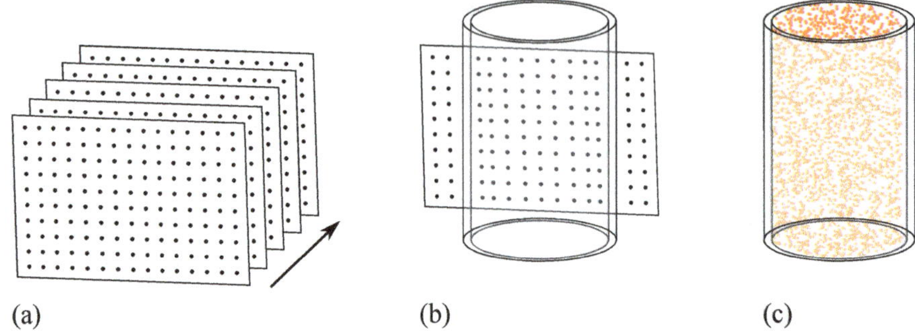

Figure 4.8: Camera calibration procedure: (a) identification of the pinhole-camera parameters via target-based calibration; (b) identification of the cylinder parameters; (c) volume self-calibration.

used to solve the full pinhole camera model with lens distortion corrections using the parameters from the DLT method as the initial values for the optimization. To improve the accuracy in the determination of the pinhole-camera parameters, a volume self-calibration can be performed still in absence of the cylinder.

Once determined the pinhole-camera parameters, the cylinder parameters can be identified by placing the cylinder in situ and recording and analyzing images of the calibration target distorted by its sidewall. This is done in principle by inserting the calibration plate inside the cylinder, placing it in several known positions within the cylinder inner volume. In the present experiments, it is not possible to operate in such a way since the cylinder is attached to the heater exchanger at its top and, thus, its interior is not accessible when it is placed in situ. As an alternative, the calibration target is placed in a known position behind the cylinder, at a small distance from the sidewall in such a way that a sufficient number of markers is in focus in that position (see figure 4.8b). Figure 4.9 reports the images of the calibration plate recorded by one of the four cameras employed in the same position with and without the cylinder placed in situ in front of the camera. From the comparison of these two images, it is possible to determine for each object point in the field of view of the camera (in absence of the cylinder) the image distortion caused by the cylinder sidewall. Indeed, the cylinder parameters are determined via nonlinear least-squares optimization starting from the known 3D world position of the markers and their 2D locations detected in the distorted image. This optimization is performed in a combined way for all the cameras of the imaging system and thus a unique set of optimal cylinder parameters for all the cameras is found, keeping the pinhole-camera parameters unchanged.

The initial values of the cylinder parameters can be easily assigned when the geometrical and physical properties of the cylindrical sample are known (or measured) and the relative location and orientation of the world reference frame with respect to the cylinder reference frame are correctly estimated. Obviously, the best practice is to choose a world reference frame such that the extrinsic cylinder parameters ($\alpha_c$, $\beta_c$, $\eta_0$, $\zeta_0$) are small, in such a way that their initial values can be set to zero. As concerns the intrinsic parameters, the internal radius and the thickness of the cylinder sidewall

 | CHAPTER 4. Camera calibration model for imaging through a cylinder

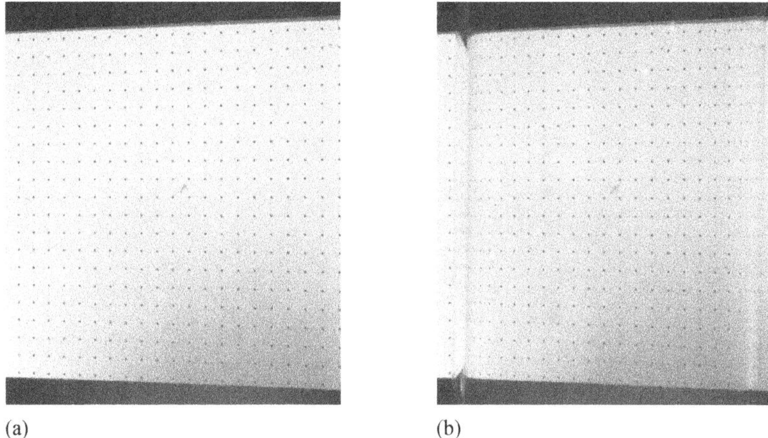

(a)         (b)

Figure 4.9: Image of the calibration plate (a) without the cylinder and (b) with the cylinder in situ.

can be measured, whereas the value of the refractive index ratios can be inferred from databases once the material constituting the sidewall and the working fluid are known. It is interesting to note that uncertainty on the latter parameters does not play a fundamental role, as proved by some tests performed both numerically and experimentally showing that convergence is obtained also starting from values of $\varrho_e$ and $\varrho_i$ different by $90\%$ from their optimized ones.

The last step of the calibration procedure consists in a volume self-calibration (figure 4.8c), where all the parameters of the model, both the pinhole-camera ones and the cylinder ones, are optimized simultaneously.

It is worth noting that the above procedure is effective when it is possible to place the cylinder in situ after the calibration of the pinhole-camera parameters. Otherwise, the estimation of the calibration parameters can be considerably hard, since the non-linearity of the model implies, as above explained, use of iterative methods, the convergence of which strongly depends on the initial guess values. While the cylinder parameters can be readily estimated, some of the pinhole camera parameters might be difficult to guess (for instance, consider that the actual focal length $f$ depends on both the lens employed and the presence of optical windows or lens filters). In the latter case, a feasible approach might be the following one. The cylinder interior has to be swept by a calibration plate, which should be designed appropriately to cover regions adjacent to the cylinder sidewall. As a first step, the pinhole-camera parameters can be estimated by using the control points affected by the smallest cylindrical ray distortions. For this purpose, it is possible to neglect the cylindrical distortion and use the DLT for estimation of the perspective projection parameters and then determine the distortion parameters from the full pinhole-camera model by means of non-linear optimization methods. This procedure is carried out separately for each camera and provides a fairly good estimation of the pinhole-camera parameters to be used as initial values for the calibration of the full

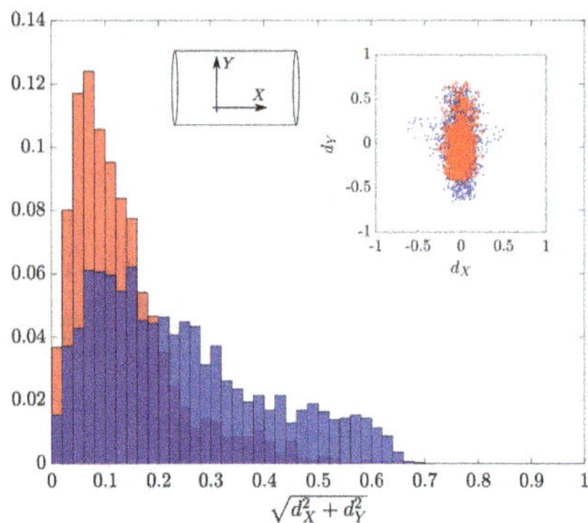


Figure 4.10: Experimental assessment of the novel pinhole-camera model with the refraction correction for the cylindrical deformation against the classical pinhole camera model. Histograms and scatter plots of the calibration residual errors. The blue bars and dots correspond to the classical pinhole camera model, the red ones to the novel model with the refraction correction. Histogram are normalized. Disparity values are in pixel units.

model with the refraction correction for the cylindrical distortion. The following step is indeed to optimize both the pinhole-camera parameters and the cylinder ones using all the control points available. Finally, in the case of PIV applications, a further refinement of the calibration parameters can be obtained via volume self-calibration.

## 4.5. Assessment of the novel model

IN the present section, the novel camera calibration model is comparatively assessed against the classical pinhole camera model (section 4.2) and the polynomial models (see section 2.4). The comparison is based on the experimental data deriving from the calibrations performed for the present T-PIV investigation of RB convection.

Figure 4.10 reports a comparison of the distributions of the self-calibration residual disparities of the classical pinhole camera model and the above-introduced model for one of the four cameras employed. Such distributions have been obtained using an identical self-calibration process (in particular, the same number of sub-volumes where to calculate the disparity errors) and starting from the same pinhole-camera calibration parameters, which were determined via a target-based calibration. The initial values of the cylinder parameters for the self-calibration were determined through the first procedure described in the previous section of this chapter. The histogram related to the camera model with the refraction correction for the cylindrical deformation (red) shows that the most frequent values of the disparity errors are around 0.08 pixels, while the relative frequency goes to zero at about 0.5 pixels. Conversely, the histogram related to the classical pinhole camera model (blue) exhibits a broader distribution with fairly high values of the relative frequency of occurrence up to 0.64 pixels. When observing the error distribution in the disparity

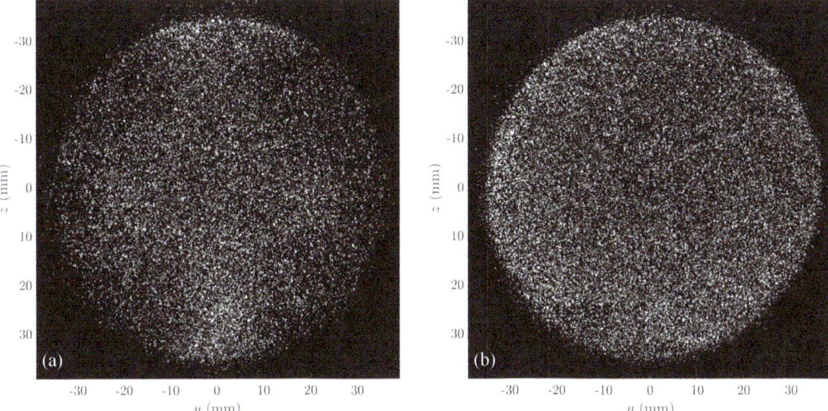

Figure 4.11: Tomographic reconstruction of the light intensity field by means of (a) the classical pinhole camera model and (b) the model with the refraction correction for the cylindrical deformation. Projection view along the direction of the cylinder axis.

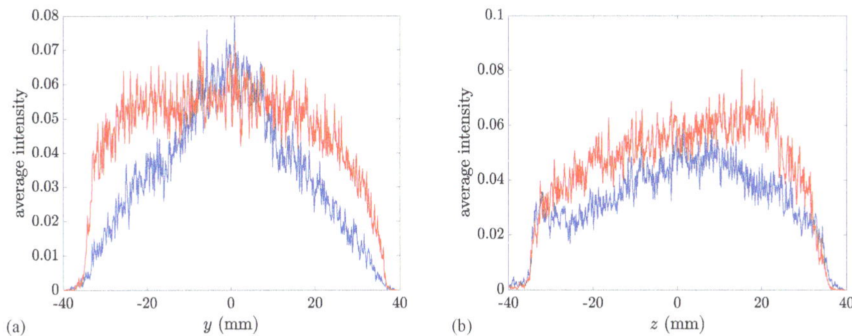

Figure 4.12: Averaged profile of the voxel intensity (a) in the $xz$-plane and (b) in the $yz$-plane corresponding to the reconstructions of the snapshot in figure 4.11. Blue curves correspond to the reconstruction based on the classical pinhole camera model, red curves to the model with the refraction correction for the cylindrical deformation.

plane (scatter plot in the top right part of figure 4.10), it is clear that the greatest component of the disparity errors is that in the $Y$-direction, which is roughly normal to the cylinder axis, as reasonably expected.

Figure 4.11 compares the reconstructions of the light intensity field related to the same snapshot obtained by means of the two pinhole camera models without and with the refraction correction for the cylindrical distortion. For such reconstructions, an SMTE procedure relying on CSMART and SMART algorithms has been used (corresponding to the first stage of the procedure described in section 3.1.4); the

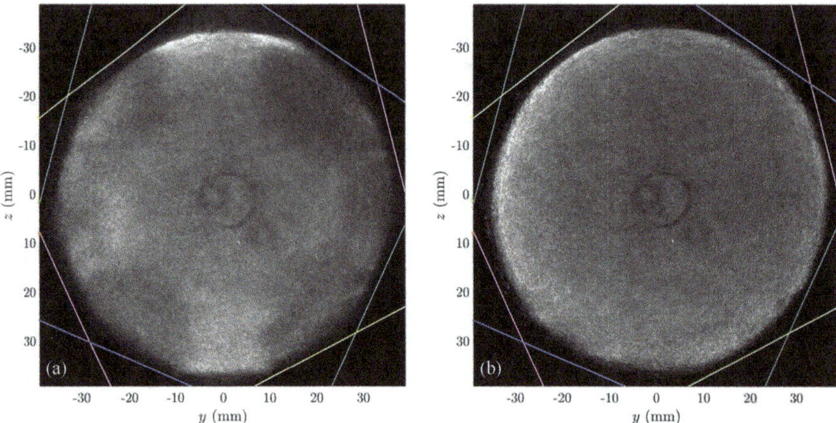

Figure 4.13: Reconstructed light intensity field averaged over 500 snapshots by means of (a) the classical pinhole camera model and (b) the model with the refraction correction for the cylindrical deformation. Projection view along the direction of the cylinder axis. Colored lines represent the LOSs of the four different cameras delimiting the region of the reconstructed volume with non-zero light intensity values in the case of the classical pinhole camera model.

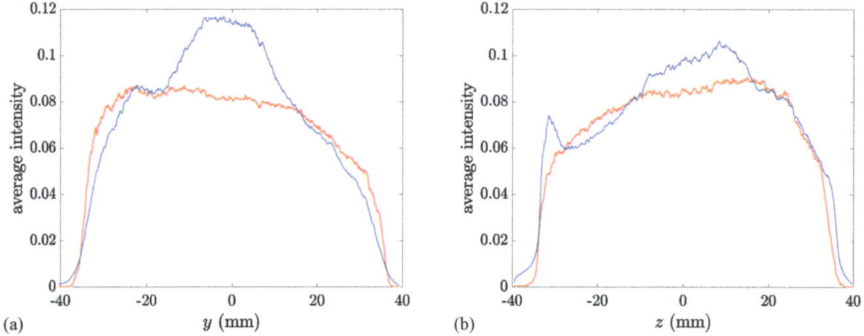

Figure 4.14: Averaged profile of the voxel intensity (a) in the $xz$-plane and (b) in the $yz$-plane corresponding to the time-averaged reconstructions of the snapshots reported in figure 4.13. Blue curves correspond to the reconstruction based on the classical pinhole camera model, red curves to the model with the refraction correction for the cylindrical deformation.

presented snapshot is the fifth reconstructed one in the sequence of the snapshots analyzed. It is remarked that the imaging system consists of four cameras lying in a plane parallel to the $yz$-plane and forming angles with the $z$-axis angles of about $-60°$, $-20°$, $20°$ and $60°$, respectively. As visible in figure 4.11, when using the pinhole camera model without the correction for the cylindrical deformation, such a configuration leads to significant losses of light intensity in the reconstructed field especially in the regions at large distances from the $xz$-plane. This is evident in figure 4.12 where the average profiles of the voxel intensity in the $xz$-plane and in the $yz$-plane corresponding to the snapshot in figure 4.11 are reported. The most significant differences are found in the $y$-profile of the laser beam (figure 4.12a): the profile corresponding to the reconstruction with the classical pinhole camera model has a maximum at the location of the cylinder axis ($y = 0$ mm) and decreases almost linearly moving away from it; on the other side, the profile related to the corrected model exhibits a nearly uniform distribution over the region covered by the laser beam. As concerns the $z$-profile, the difference in the light intensity is distributed over the entire depth of the laser beam.

Figure 4.13 reports the reconstruction of the light intensity field averaged over 500 snapshots (corresponding to about 1 min of measurements), while figure 4.14 shows the corresponding average profiles in the $xz$- and $yz$-planes. Differently from the results of figure 4.11, these reconstructions are based on the full process described in section 3.1.4. Therefore, all the snapshots, except the first 10, are analyzed via STB and IPR and the corresponding 3D light fields are determined starting from particles and using an appropriate imaging model. More specifically, each particle is treated as an isotropic Gaussian light blob and is identified by a set of five parameters corresponding to its 3D world coordinates, its diameter and the maximum light intensity level of the blob [176].

In figure 4.13a, it is possible to note that the use of the classical pinhole camera model leads to the generation of a considerable amount of ghost light outside of the region actually illuminated. This is clearly shown by the colored lines superimposed to the picture, which represent the LOSs of the four different cameras (projected in the considered view) delimiting the region of the analyzed volume with non-zero light intensity values. In particular, ghost particles are concentrated in the region around $z = -30$ mm, i.e. the backward part of the cylinder interior, as a consequence of the fact that the LOSs along which all the cameras image particles in such a portion of the measurement domain intersect the latter for a fairly large depth. This behavior results in a sort of "bump" extending over $-15$ mm $< y < 15$ mm in the $y$-profile of the average light intensity field (see blue curve in figure 4.14a). Outside of such a range, the average intensity in the $xz$-plane for the case of the classical pinhole camera model is lower than that corresponding to the reconstruction with the corrected model (red curve) up to the edge of the laser beam (approximately the cylinder sidewall), whereas beyond this point we observe a ghost intensity level in the first case and zero values in the second one. In fact, very few artifacts are found in the reconstruction with the corrected model as also noticeable in figure 4.13b. As already seen for the snapshots in figure 4.12, the differences in the $z$-profiles (figure 4.14b) are less relevant, although in the case of the classical pinhole camera model ghost particles cause larger levels of the average light intensity in the central region and a pronounced local peak at the edge of the light beam around $z = -37$ mm. It should also be noted that in both the time-averaged reconstructions of figure 4.13 a small ring-like region of lower intensity level is found at a radial distance from the cylinder axis approximately equal to 5 mm. The latter corresponds to the edge of the central hole of the layer C of the heat exchanger shown in figure 3.4, which is evidently the only part of the top cooling system that is not perfectly transparent.

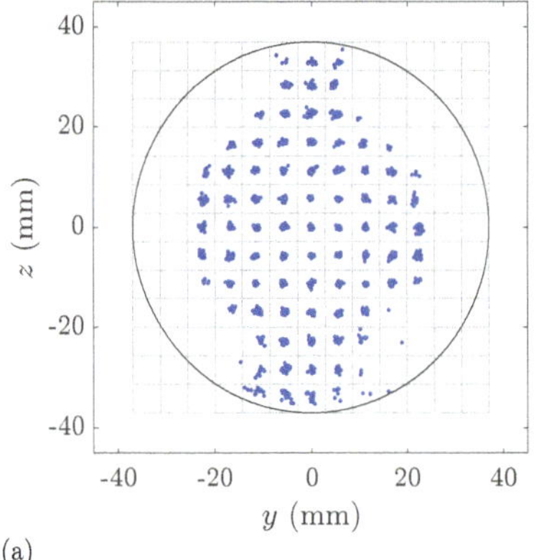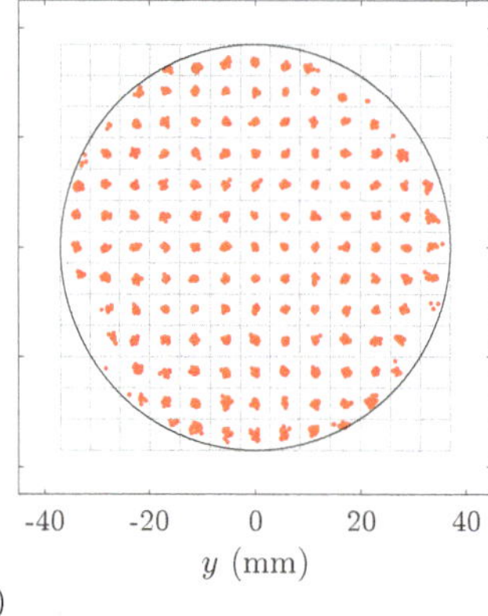

Figure 4.15: Volume self-calibration with (a) the classical pinhole camera model and (b) the model with the refraction correction for the cylindrical deformation. Mean positions of the triangulated particles falling in separate sub-volumes of the whole illuminated volume. The sub-volumes are represented with solid gray lines.

The above analysis suggests that the classical pinhole camera model does not provide sufficient accuracy for the tomographic reconstruction in the regions adjacent to the cylinder sidewall. This seems to be in disagreement with figure 4.10 where maximum disparity errors for the classical pinhole camera model are found to be lower than 0.6 pixels. The reason of this apparent inconsistency is that in the self-calibration procedure it is not possible to assign a disparity vector to sub-volumes where the triangulation errors are too large. In fact, in such sub-volumes the disparity maps are characterized by a large dispersion (i.e., the sum of the disparity vectors corresponding to the matched particles is not convergent) and it is not possible to detect a principal peak with sufficient signal-to-noise (SN) ratio[1]. In order to assess the accuracy of the camera calibration model over the different regions of the illuminated volume, figure 4.15 reports the mean positions of the triangulated particles in the sub-volumes where a disparity vector is detected with sufficient SN ratio (greater than 2) for both the case of the classical pinhole camera model and the novel model. The illuminated volume is divided in $26 \times 13 \times 13$ sub-volumes with the same dimensions along the three coordinate directions (about 80 voxels corresponding to 5.6 mm). The figure confirms that the regions adjacent to the cylinder sidewall at large $y$ locations are affected by the greatest triangulation errors; in particular, these errors are expected to be greater than the search radius used in the triangulation procedure itself, which is equal to 2 pixels for the considered case.

All the previous observations are an experimental evidence that the pinhole camera model without any modification is not suitable for mapping of the cylinder

---

[1] The signal-to-noise ratio is defined as the ratio of the principal peak to the second largest peak detected in the disparity map.

 CHAPTER 4. Camera calibration model for imaging through a cylinder

interior with the accuracy required for T-PIV measurements. In the following, we try to explain the superior performance of the corrected pinhole camera model in the comparison with the polynomial mapping functions. As aforementioned, this model has the advantage that there is no need of sweeping a calibration target through the cylinder interior to perform the camera calibration. On the other hand, this is not possible in the case of polynomial camera models for two main reasons: first, polynomials are smooth functions and, as such, they cannot generally describe correctly the projection transformation both outside and inside the cylinder; second, it is well known that extrapolation via polynomials can lead to significant triangulation errors. Therefore, in order to correctly calibrate a polynomial camera model, an appropriate calibration target should be designed in such a way to place control points everywhere in the cylinder interior, even sufficiently near the sidewall. In the present experiments, a direct comparison between polynomials and the innovative camera model is not possible due to inaccessibility of the cylinder volume, which makes calibration of the polynomial models impracticable. However, a virtual calibration of the polynomial models can be performed assuming that the correspondence between the 3D world and 2D image coordinates is exactly described by the pinhole camera model with the refraction correction for the cylindrical distortion. Figure 4.16 reports the results of the virtual calibrations of different polynomial models in terms of the distribution of the calibration residual errors. These calibrations are based on a grid of $28 \times 14 \times 14$ control points with constant spacing along the three coordinate directions equal to about $5$ mm (70 voxels); indeed, only points falling in the cylinder interior are used.

For completeness, the results of the virtual calibration for the classical pinhole camera are reported in figure 4.16a. Differently from the counterpart of figure 4.10, the error distribution in figure 4.16a is characterized by a greater dispersion. This is due to the fact that the latter includes errors related points adjacent to the cylinder sidewall, which contribute to the estimation of the calibration parameters in the same way as points located in the central part of the cylinder inner volume. In figure 4.16b three different types of polynomial functions are compared: 3rd, 5th and 6th-order, respectively. It can be noted that the 3rd-order polynomial function leads to unacceptably high errors, while the other two models offers better performance, with the higher order polynomial having about the $75\%$ of errors below 0.1 pixels. Figures 4.16c-d present the distribution of the calibration errors for several multi-plane polynomial models using a different order of in-plane polynomials (third order in figure 4.16c and fifth order in Figure 4.16d) and a different number of $z$-planes (2, 6 and 9 respectively). It is worth noting that, in the current implementation of the multi-plane polynomial model, the number of the base $z$-planes to be used for the piecewise linear interpolation in the $z$ direction is chosen independently from the distribution of the control points. This means that it is possible to select a base $z$-plane that contains no control points; the coefficients of the corresponding polynomial are determined via linear least-squares optimization algorithms starting from the control points falling in the two $z$-intervals at the sides of the $z$-plane itself. In such a way, polynomial models using only 2 $z$-planes are meaningful, although, as shown in figures 4.16c-d, they lead to considerably large errors. Among all the

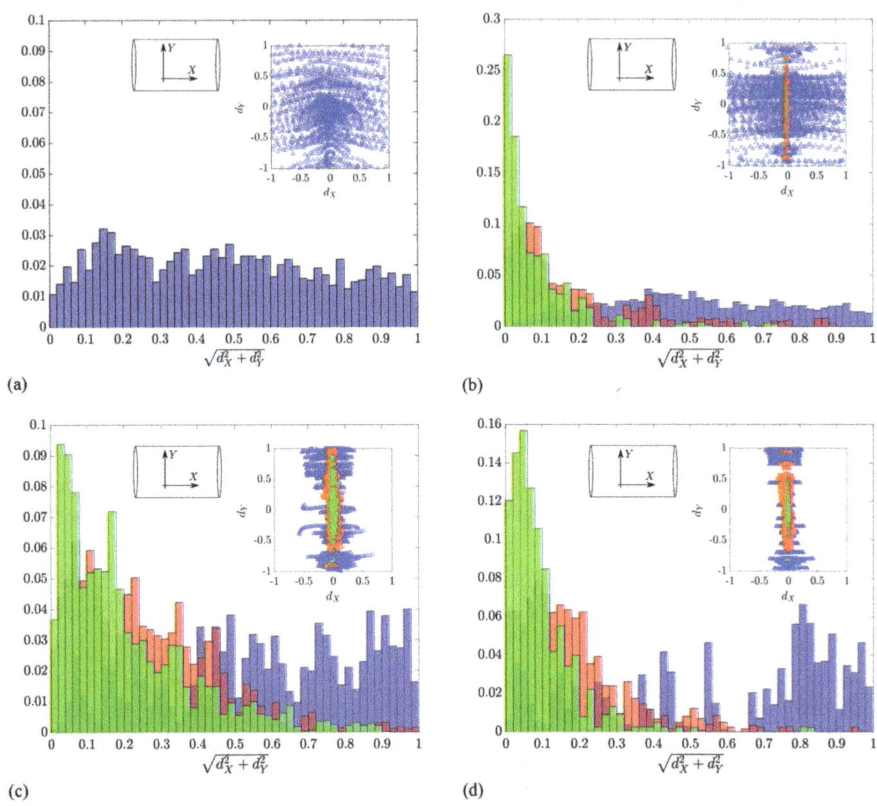

Figure 4.16: Experimental assessment of the novel pinhole-camera model with the refraction correction for the cylindrical deformation against the classical pinhole camera model and polynomial models. Histograms and scatter plots of the calibration residual errors. Results of the virtual camera calibrations performed using: (a) the classical pinhole camera model; (b) polynomial mapping functions with orders: $N_x = N_y = N_z = 3$ (blue), $N_x = N_y = N_z = 5$ (red), $N_x = N_y = N_z = 6$ (green); (c) multi-plane polynomial models based on third-order polynomials on selected $z$-planes and piecewise linear interpolations along the $z$-direction; the numbers of $z$-planes employed are 2 (blue), 6 (red) and 9 (green). (d) multi-plane polynomial models based on fifth-order polynomials on selected $z$-planes and piecewise linear interpolations along the $z$-direction; the numbers of $z$-planes employed are 2 (blue), 6 (red) and 9 (green). Histogram are normalized. Disparity values are in pixel units.

investigated multi-plane polynomial models, that using fifth order polynomials on 9 nine different $z$ planes appears to have a performance comparable with that of the modified pinhole camera model. However, there are at least two problems related to the use of polynomial functions: introduction of local oscillations with increasing the order of the polynomial and the very large number of the calibration parameters involved.

Figure 4.17 shows the phenomenon of the oscillations introduced by high order polynomial function. Here, $\Delta Y$ represents the difference of the $Y$ coordinate of an object point projected by using a polynomial model and the $Y$ coordinate of the same

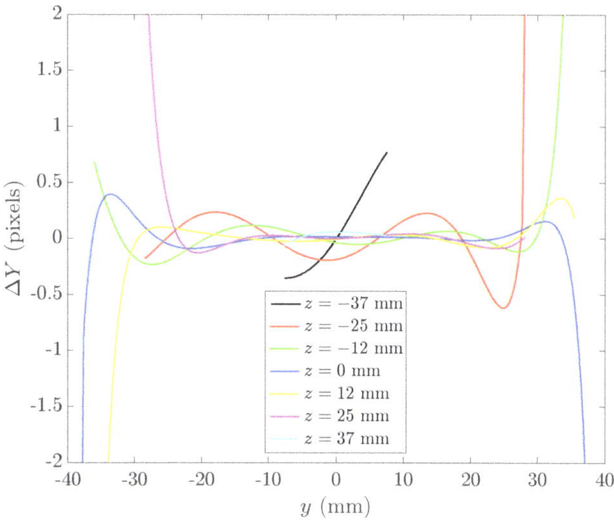

Figure 4.17: Local projection error of the polynomial model with 6th order in $x$, $y$ and $z$ at fixed $x$ location and variable $z$ location. The errors are relative to the results of the pinhole-camera model with the refraction correction for the cylindrical distortion, assumed as representative of the exact correspondence between the 3D world and the 2D image coordinates.

point projected via the pinhole camera model with the refraction correction for the cylindrical distortion. The polynomial function considered is the 6th-order one and the behaviors of $\Delta Y$ along chords parallel to the $y$-axis and contained in the $yz$-plane for different $z$ locations are shown. From the figure it is evident that in the central part of the cylinder volume the image positions obtained by the polynomial model oscillate around the exact solution (assumed to be represented by the corrected pinhole camera model) of up to 0.2 pixels. These oscillations result in an augmented noise in the tomographic reconstruction or iterative particle identification, increasing the uncertainty in the particle displacement estimation. Moreover, the error in proximity of the cylinder sidewall is very large (greater than 2 pixels in some cases) and this suggests that a very high order is indeed required to correctly map these regions of the cylinder interior. Finally, the number of the calibration parameters in the case of the 6th-order polynomial function is 168 against the 25 of the pinhole camera model with the refraction correction for the cylindrical distortion.

In conclusion, table 4.1 summarizes the main features of the camera models investigated in this paragraph. From the table it is possible to understand that the polynomial models that offer errors comparable to those of the pinhole camera model with the refraction correction for the cylindrical distortion (polynomial with 6th order in $x$, $y$ and $z$ and the multi-plane polynomial with 5th order and 9 planes) are characterized by a number of calibration constants more than 6 times greater (168 and 378, respectively, against 25).

Table 4.1: Main features of the investigated camera models: number of parameters, root mean square error, median error and 99th percentile of the calibration residual error distribution. Results are related to the virtual calibrations performed starting from the calibrated pinhole camera model with the refraction correction. Results for the reference model are related to volume self-calibration. Errors are in pixel units.

| model | # of par. | rms err. | median err. | 99th percent. |
|---|---|---|---|---|
| pinhole with correction[*] | 25 | 0.13 | 0.11 | 0.47 |
| classical pinhole | 17 | 1.42 | 1.06 | 6.37 |
| polyn. $3 \times 3 \times 3$ | 40 | 0.53 | 0.36 | 3.72 |
| polyn. $5 \times 5 \times 5$ | 112 | 0.18 | 0.10 | 1.05 |
| polyn. $6 \times 6 \times 6$ | 168 | 0.085 | 0.047 | 0.65 |
| 2 $z$-planes polyn. $3 \times 3$[†] | 40 | 3.48 | 3.27 | 10.27 |
| 6 $z$-planes polyn. $3 \times 3$[†] | 120 | 0.29 | 0.23 | 1.34 |
| 9 $z$-planes polyn. $3 \times 3$[†] | 180 | 0.23 | 0.16 | 1.19 |
| 2 $z$-planes polyn. $5 \times 5$[†] | 84 | 3.46 | 3.30 | 10.54 |
| 6 $z$-planes polyn. $5 \times 5$[†] | 252 | 0.19 | 0.15 | 0.76 |
| 9 $z$-planes polyn. $5 \times 5$[†] | 378 | 0.097 | 0.07 | 0.44 |

[*] reference model.
[†] piecewise linear interpolation along the $z$-direction.

**5**

# Investigation of non-rotating Rayleigh-Bénard convection

T HIS chapter focuses on the results of the T-PIV investigation of RB convection in the cylindrical sample with aspect ratio $\Gamma = 1/2$ at Rayleigh number equal to $1.86 \times 10^8$ and Prandtl number equal to 7.6. The values of these governing parameters are based on the physical and thermal properties of the water at the nominal mean temperature between the top and the bottom of the convective cell.

In the present experiment, the temperature of the bottom is $20\,°C$ and the temperature of the top is $15\,°C$, thus the nominal mean temperature is $17.5\,°C$, while the nominal temperature difference is $5\,°C$. The actual temperature difference over the fluid is reduced by the temperature drop across the Plexiglas top plate. Such a drop and the effective Rayleigh number can be calculated by means of a one-dimensional heat conduction analysis; this requires the knowledge of the Nusselt number of the convective system in order to define its effective thermal conductivity. Preliminary simulations were performed to estimate the Nusselt number at the nominal operating Rayleigh number. Based on the results of the simulations, the estimated temperature drop across the Plexiglas plate is $0.048\,°C$, which results in an effective mean temperature of $17.476\,°C$. The Prandtl number and the other properties of the working fluid at this temperature are practically coincident with those at the nominal mean temperature.

Once set the temperature difference, the system is allowed to adapt for about two hours before starting the experiment. Time-resolved measurements have been carried out over four hours with a sampling frequency of 7.5 Hz (corresponding to a sequence of 108,000 snapshots). A subset of the data corresponding to two hours has been processed in the whole with the procedure described in section 3.1.4. To improve statistical convergence, one instantaneous velocity field per each minute of the remaining two hours has also been used in the computation of the statistical fields.

# 5.1. Structure of the mean velocity field and its relationship with the instantaneous evolution

FIGURE 5.1 shows the morphology of the time-averaged velocity field. The time-average velocity field is found to be azimuthally axisymmetric, as observed in previous experimental and numerical investigation [45, 57, 193]. Therefore, the structure of the mean flow can be inspected by focusing on the dynamics in one of the azimuthal planes (passing through the cylinder axis). For this purpose, the contour map of the vertical velocity component (component along the cylinder axis $x$) in the $xz$-plane is reported in figure 5.1a; here, velocity values are scaled by the so-called free-fall velocity $u_0$ which in the present case is equal to 38.2 mm/s (assuming as $\Delta$ the effective temperature difference over the fluid). In the lower part of the cylinder, the flow is observed to rise up in the regions adjacent to the sidewalls and fall down in the central part; a specular behavior is observed in the upper part. This results in four rolls in the planar pattern (shown clearly by the velocity vectors superimposed to the map in figure 5.1a), which indeed are related to the existence of two toroidal vortex structures near the top and bottom plates. These vortex structures have been identified in the 3D flow field by the $Q$-criterion [65, 194], i.e. as the regions of the flow field where the second invariant of the velocity gradient tensor is positive and greater than a predefined threshold, and represented in figure 5.1b, along with the isosurfaces of the vertical velocity component corresponding to values of $\pm 0.026\,u_0$ (red and blue for positive and negative, respectively).

This structure of the flow field is strictly connected to the dynamics of the plumes and the LSC. In fact, the region of positive (negative) vertical velocities adjacent to the cylinder sidewall in the lower (upper) half of the cell is related to hot (cold) plumes detaching from the thermal boundary layer on the bottom (top) plate and rising upward (falling down); conversely, the central region of negative (positive) velocities is associated with the arrival of the plumes coming from the opposite plate. Such dynamics is clear in figure 5.2, which reports two instantaneous velocity fields corresponding to two different times of the measurements. In the instantaneous velocity, the axisymmetry is broken and the flow organizes in ascending and descending currents which extend over the entire cell height. Such currents are indeed clusters of plumes that survive the shearing and the mixing of the turbulent flow in the bulk and reach the opposite plate, as noted by Sun *et al.* [123]. The clusters of plumes are observed to deviate from the sidewall at a certain height as a consequence of the thermal and kinematic diffusion experienced during their rising/falling and finally impinge the opposite plate in the central part. In both the instantaneous velocity fields reported in figure 5.2 the vertical currents form a LSC, which exhibits an elliptical structure inclined with respect to the cylinder axis with two counterrotating rolls in the corners where the currents deviate from the sidewall.

It is worth noting that in the selected snapshots the LSC is localized in two perpendicular planes (the $xy$ and $xz$ coordinate planes). As discussed in the introduction of this dissertation, due to the axisymmetry of the geometry and the boundary conditions, in a cylindrical sample there may exist no preferential orientation of the LSC. In fact, all the possible azimuthal orientations are observed in the present

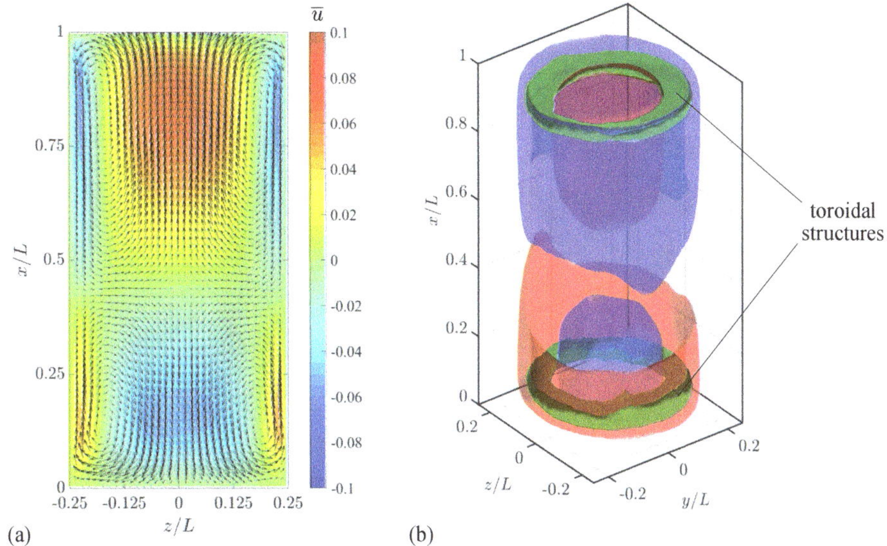

Figure 5.1: Mean velocity flow field: (a) vertical velocity component map with vectors superimposed (velocity are scaled by the free fall velocity $u_0$); (b) isosurfaces of the vertical velocity component corresponding to values of $\pm\,0.026u_0$ (red and blue for the positive and negative values, respectively) and isosurfaces of the second invariant of the velocity gradient field showing the mean vortical structures, corresponding to approximately 10% of the maximum value.

experiment. Sun *et al.* [45] identified the continual reorientation of the elliptic LSC as responsible of the azimuthal symmetry and the existence of the vortex-ring-like structures attached to the bottom and the top plates in the time-average velocity fields. Such a picture is simplistic in the range of Rayleigh and Prandtl numbers including the investigated case, since the dynamics of the convective flow is characterized by a variety of different states and transitions from one state to another. A more detailed description of the flow states and their statistical occurrence is presented in the following sections. However, figure 5.3 reports a snapshot of the velocity field in which the dominant pattern is not a single-roll structure, but indeed the flow appears to be organized in two counterrotating vertically stacked rolls, laying approximately in the $xy$-plane, each extending over one half of the convective cell. Such a state is typically referred to as "double-roll state" (DRS) in opposition to the "single-roll state" (SRS) which essentially consists of the domain-filling LSC. The existence of the DRS in a cylinder with $\Gamma = 1/2$ was first found numerically by Verzicco and Camussi [57] and then observed in several experimental investigations, among which are the investigations of Xia *et al.* [92] and Weiss and Ahlers [195]. Obviously, also the DRS has no preferential orientation and thus its occurrence does not introduce azimuthal asymmetries in the time-averaged velocity field.

From figure 5.1 it can be also seen that the time-averaged velocity field is characterized by the presence of a saddle point located on the cylinder axis. One could expect the saddle point to be positioned at half-height of the cell, while in the present

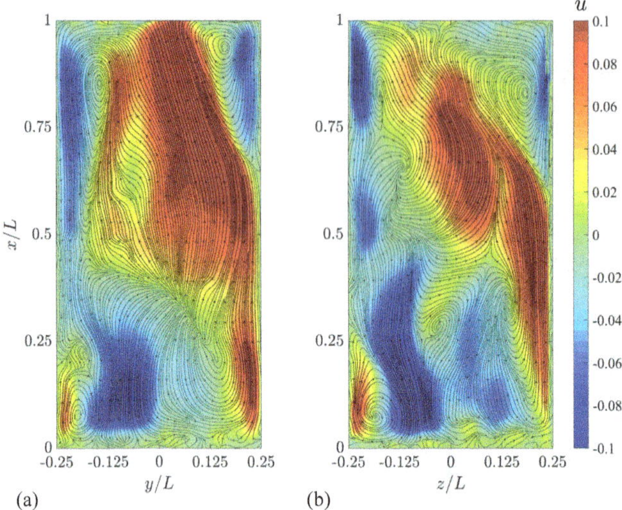


Figure 5.2: Instantaneous maps of the vertical velocity component (a) in the $xy$-plane and (b) in the $xz$-plane at two different times. Superimposed are the 2D streamlines. Values are scaled by the free fall velocity $u_0$.

case it is found at $x/D \approx 0.43$. This is due to fact that the cylinder sidewall is not adiabatic and, in the present experiment, the temperature of its external side (i.e., the tank temperature) was lower than the average temperature over the cylinder volume. More specifically, the tank temperature was not controlled by means of the corresponding cooling system (see chapter 3.1.1) and set to a specific value. It was simply measured and found to be equal to $17.35\,^{\circ}\mathrm{C}$.

The effects of the temperature at the external side of the cylinder sidewall on the time-average velocity field are shown in figure 5.4, where the results of the simulations of the RB convection in a one-half aspect ratio cylinder with a physical sidewall and isothermal boundary condition at its external side are reported for different values of the imposed temperature $T_e$. The Prandtl and the Rayleigh numbers are the same of the experimental case, as well as the physical properties of the sidewall ($(\rho C_p)_s/(\rho C_p)_f = 0.5$, $\kappa_s/\kappa_f = 0.32$ and $t_s/L = 0.0207$). In figure 5.4, the time-mean velocity fields have been averaged azimuthally by exploiting the axisymmetry of the flow field.

Figure 5.4 shows that, by decreasing the temperature $T_e$ from the value $T_m$ corresponding to the mean temperature of the fluid to the temperature $T_b$ of the bottom plate, the axisymmetric radial recirculation in the lower part of the cell reduces progressively, while that in the upper part grows and covers a larger and larger portion of the sample. This corresponds to a strengthening of the toroidal structure attached to the top plate and a weakining of the one near the bottom plate. Such an average behavior can be explained by considering that when $T_e < T_m$ the hot plumes moving near the sidewall lose their heat excess faster than the cold plumes, which are subjected to a reduced heat transfer with the physical sidewall,

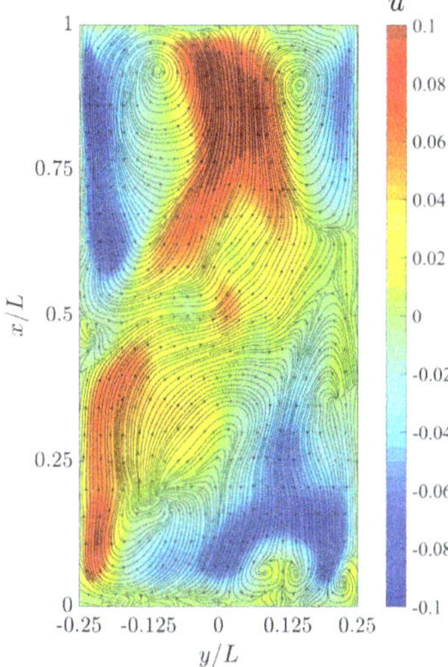


Figure 5.3: Snapshot of the velocity field in the double-roll state: maps of the vertical velocity component with 2D streamlines superimposed. Values are scaled by the free fall velocity $u_0$.

and thus they are carried into the bulk motion earlier (i.e., at a smaller distance from the bottom plate) with respect to the case $T_e = T_m$ or the case with adiabatic wall. In the extreme case $T_e = T_b$ hot plumes detaching near the sidewall are rapidly pushed away from it by the descending opposite cold currents and this results in the formation of a very small vortex near the lower corner of the cylinder, which is the only legacy of the lower toroidal structure observed in the case $T_e > T_b$ (see figure 5.4). Consequently, the average flow consists of a unique domain-filling axisymmetric radial recirculation with ascending fluid in the center and descending fluid near the sidewall. Obviously, an increase of $T_e$ above the mean temperature $T_m$ would cause an opposite evolution, with the recirculation in lower part of the cell growing at the expenses of that in the upper part.

The experimental measurements presented in this chapter correspond to the case presented in figure 5.4b, that is $\theta_e = (T_e - T_b)/\Delta = 0.48$. By visual inspection, it is possible to note that the structure of mean flow field is considerably similar between the experiment and the simulation and, in particular, the saddle point is found at the same height in both cases. Figure 5.5 presents a quantitative comparison of the experimental and numerical data, by reporting the vertical profiles of the time- and azimuthally-averaged vertical velocity at different radial locations. The agreement between the measurements and the numerical results is very good for $r/L \leq 0.6$, i.e. in the central part of the sample, whereas slightly greater differences are observed moving towards the sidewall. In figure 5.6 the experimental and numerical maps of the azimuthally-averaged root-mean-square vertical velocity fluctuation are comparatively shown. Although the spatial distributions look identical, different levels of fluctuation are observed between the experiment and the simulation. In

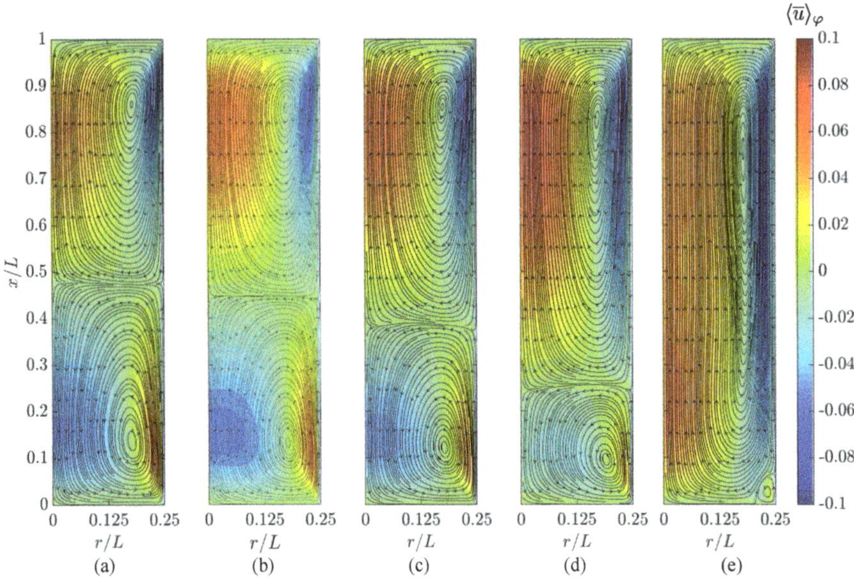


Figure 5.4: Effects of the temperature at the external side of the cylinder sidewall on the time-average velocity field. Results of the simulations at $Ra = 1.86 \times 10^8$, $Pr = 7.6$ and $\Gamma = 0.5$, including the presence of a physical sidewall with the same thermal and geometrical properties of that of the present experimental apparatus $((\rho C_p)_s/(\rho C_p)_f = 0.5$, $\kappa_s/\kappa_f = 0.32$ and $t_s/L = 0.0207)$. The boundary condition at the external side of the wall is isothermal and the imposed temperature is: (a) $T_e = T_b + 0.50\Delta = T_m$ $(\theta_e = 0.50)$; (b) $T_e = T_b + 0.48\Delta$ $(\theta_e = 0.48)$; (c) $T_e = T_b + 0.46\Delta$ $(\theta_e = 0.46)$; (d) $T_e = T_b + 0.40\Delta$ $(\theta_e = 0.40)$; (e) $T_e = T_b$ $(\theta_e = 0)$. Contour maps of the time- and azimuthally-averaged axial velocity component with streamlines superimposed. Values are scaled by the free fall velocity $u_0$.

particular, the fluctuations measured experimentally have smaller values than those computed numerically. Indeed, the maximum vertical velocity fluctuation intensity is found to be $0.054\,u_0$ in the experimental case and $0.065\,u_0$ in the numerical one, which amounts to a percentage difference of 17%. It is very difficult to argue about the origin of such a discrepancy, which might be related to some specific feature of the experimental apparatus that is not correctly (or at all) modeled in the numerical simulation. Among these could be the geometry of the connection between the sidewall and the plates not correctly modeled in the simulation, the presence of a temperature gradient in the tank (which would imply a non-uniform surface distribution of $T_e$) and the modulation of the Lagrangian velocity measurements due to the interpolation on a structured grid in the T-PIV process.

It should however be considered that the relevance of one of the first two factors could involve significant asymmetries in the time-averaged velocity field, which, in fact, are not observed. On the other side, in order to assess the importance of modulation effects in the experimental determination of the fluctuating velocity fields, data obtained by the numerical simulations has been reduced with a process analogous to that used for the T-PIV measurements, based on a sparse sampling of the instantaneous velocity fields and reinterpolation onto the computational grid.

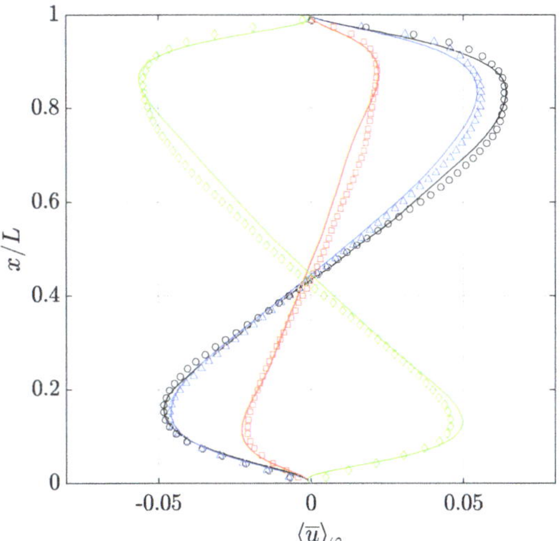

Figure 5.5: Comparison between experimental measurements and numerical results. Vertical profiles of the time- and azimuthally-averaged vertical velocity at different radial locations: $r/L = 0$, black; $r/L = 0.3$, blue; $r/L = 0.6$, red; $r/L = 0.9$, green. The symbols represent the experimental data, the solid lines the numerical solutions.

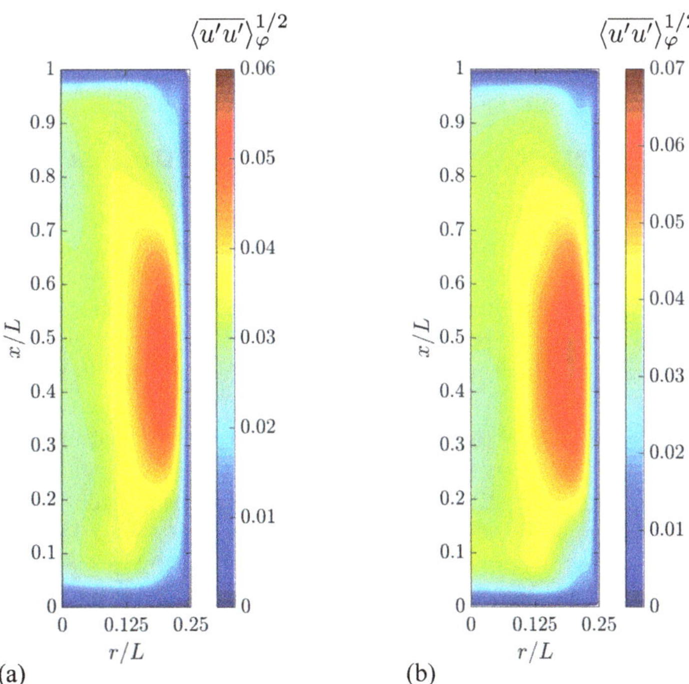

Figure 5.6: Maps of the azimuthally-averaged root-mean-square vertical velocity fluctuation from (a) experimental measurements and (b) numerical simulation. Values are scaled by the free fall velocity $u_0$.

5.1. Structure of the mean velocity field and its relationship with the instantaneous evolution | 83

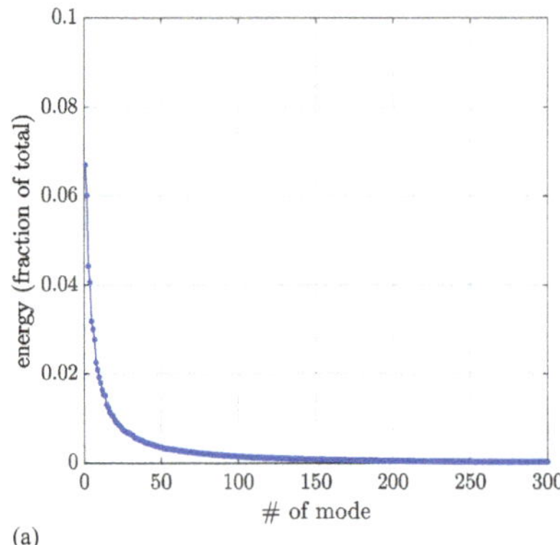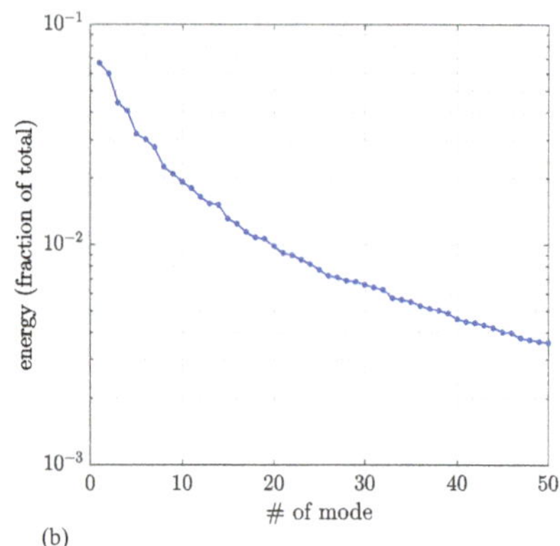

Figure 5.7: POD spectrum of the fluctuating velocity field: (a) energy of the first 200 modes normalized by the total amount of energy. (b) focus on the behavior of the normalized energy for the first 50 modes in logarithm scale.

Such process is applied to 1,000 evenly spaced numerical snapshots covering a total interval of 2,000 non-dimensional time units. Then, the root-mean-square velocity fluctuations are computed and compared with the raw numerical data. Percentage differences have been found to be lower than 1%: this means that the modulation of the velocity measurements by post-processing cannot justify the discrepancy between the experiment and the DNS. Further work is indeed required to clarify such a point.

## 5.2. Characteristic modes of the turbulent convection

IN this section, the proper orthogonal decomposition (POD) [196] is used to capture the most energetic modes of the turbulent convection. The principle and mathematical foundation of POD are outlined in appendix A. The focus here is essentially on the first POD modes which are related to the different states of the turbulent convection and the complex dynamics of the LSC, indicating the occurrence of oscillatory modes that have already been observed in literature both experimentally and numerically.

The POD is applied only to the fluctuating part of the velocity field. Figure 5.7 reports the normalized energy spectrum of the POD modes for the first 200 modes. It can be noted that the energetic budget of the first modes is not very large compared to that of the other modes. The 1st mode retains only the 6.7% of the total turbulent kinetic energy, while the sum of the energies of the first 200 modes amount to the 87% and the energy of the 200th mode is the 0.0079% of that of the 1st mode. This is a clear indication of the absence of dominant modes in the turbulent flow evolution. At a closer look, in figure 5.7b it is possible to see a sort of pairing between the normalized energy levels of the 1st and the 2nd mode (respectively equal to 6.7% and 6%), as well as those of the 3rd and the 4th mode (4.4% and 4%) and those of the 5th and the 6th mode (3.2% and 3%). This is not casual since each of these pairs

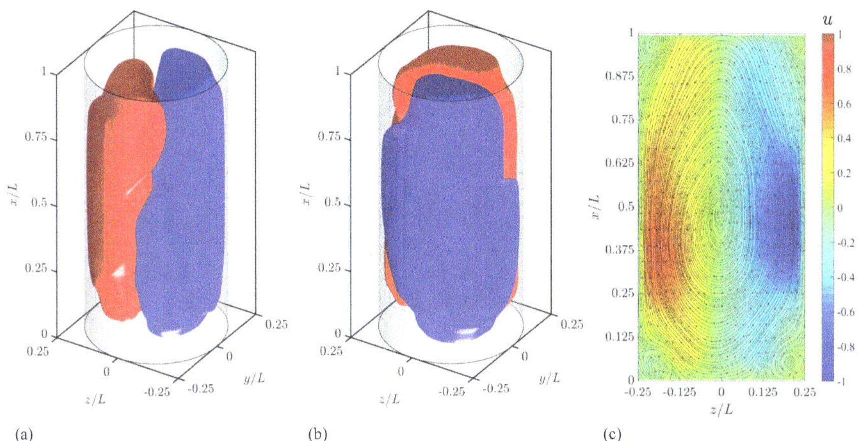

Figure 5.8: First pair of POD modes. 3D isosurfaces of the vertical velocity component corresponding to values of $\pm 10\%$ of the maximum for (a) the 1st and (b) the 2nd mode (red corresponds to the positive value) and (c) contour map of the vertical velocity component in the $xz$-plane for the 2nd mode. Values are scaled by the maximum of the vertical velocity component.

of POD modes exhibit essentially the same flow structure except for a $90°$ rotation about the cylinder axis, as shown below, and their combination allows such a pattern to occur in any possible azimuthal orientation in the instantaneous flow field.

Figure 5.8 shows the flow morphology of the first two modes via the isosurfaces of the vertical velocity component. In both modes, we note two vertical currents forming a domain-filling circulation; this is clearer in figure 5.8c, where the map of the vertical velocity in the $xz$-plane with the 2D streamlines superimposed for the 2nd mode is reported. Therefore, these POD modes are representative of the SRS observed in numerous experimental and numerical investigations over different ranges of the control parameters. It is noted that the single roll (SR) has a quasi-planar structure and is not tilted with respect to the cylinder axis; moreover, four small counterrotating rolls are observed in the corners between the SR and the plates.

This pattern has to be considered as the main contribution to the LSC existing in the instantaneous evolution. A remarkable difference between the SR of the first two modes and the LSC is that the latter has a significant inclination with respect to the cylinder axis, as visible in figure 5.2 and also observed in other works ([45] to cite one). Indeed, the inclination of the LSC results from the combination of the mean velocity field (see figure 5.1) and the first two modes. In fact, when the SR structure is superimposed to the time-average flow field, the ascending currents near the sidewall are strengthened on the side of the ascending flow in the SR and weakened on the opposite side; vice versa for the mean descending currents. This behavior leads two of the four rolls observed in 5.1a to merge in the plane of the SR structure and form the LSC, while the other two originate counterrotating rolls at the diagonally opposite corners, as observed experimentally by Sun *et al.* [45]. It is also interesting to remark that the flow far the SR plane is not significantly affected by its dynamics, so, after the superimposition of the SR structure, the pattern in the plane

5.2. Characteristic modes of the turbulent convection   |   85

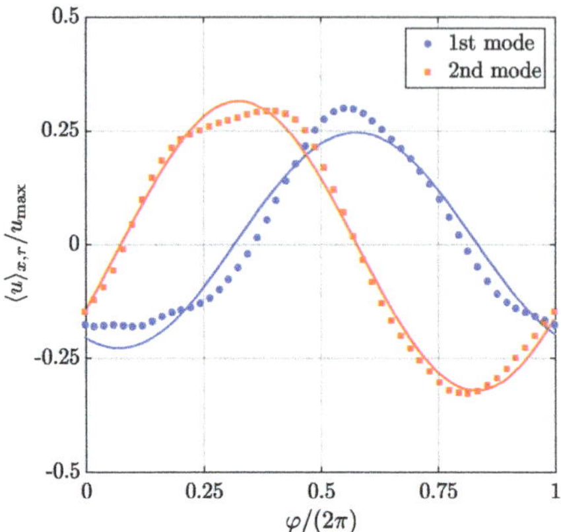

Figure 5.9: Azimuthal profiles of the axially and radially-averaged vertical velocity for the 1st and the 2nd mode. The data points represent the experimental measurements, whereas the solid lines are the cosine fits based on this probe data. Values are scaled by the maximum of the vertical velocity component in each mode.

normal to the SR plane is similar to that reported in 5.1a. This seems consistent with the observations of Sun *et al.* [95].

As aforementioned, the only significant difference between the patterns of the first two modes is their azimuthal orientation, which differs by 90°. In order to show that, figure 5.9 reports the azimuthal profiles of the axially and radially-averaged vertical velocity for both modes; the calculated phase shift between these two profiles is 91.86°. Such a feature can be explained by considering that the 1st mode could describe by itself only one azimuthal orientation of the LSC. On the contrary, the combination of the first two modes can result practically in any azimuthal orientation of the LSC plane in the instantaneous flow field. This explains also why the two modes occur to have comparable energetic levels, as discussed above.

The 3rd and 4th modes are represented in figure 5.10. Similarly to the first pair of POD modes, these modes share an identical pattern, except for a 90° rotation about the cylinder axis. Such a pattern consists of two counterrotating large-scale rolls laying approximately in the same azimuthal plane, as clearly detectable from the 2D streamlines of the 4th mode in the $xy$-plane (figure 5.10c). These POD modes resemble considerably the DRS observed in cylindrical samples with aspect ratio lower than 1 [57, 92, 195]. In the previous section, it has been showed that such a pattern is clearly detectable at certain time instants of our experiment (figure 5.3). However, it is here essential to remark that both the pairs of modes 1-2 and 3-4 generally contribute to the structure of the instantaneous velocity fields. Indeed, the emergence of the DRS in the instantaneous evolution is associated with a weak correlation of the instantaneous velocity field with the first POD modes (i.e., a small value of the projection of the snapshot onto these POD modes). On the other side, when both the contributions of the first and the second pair of POD modes are not negligible, the SRS and DRS are combined with each other in the instantaneous flow

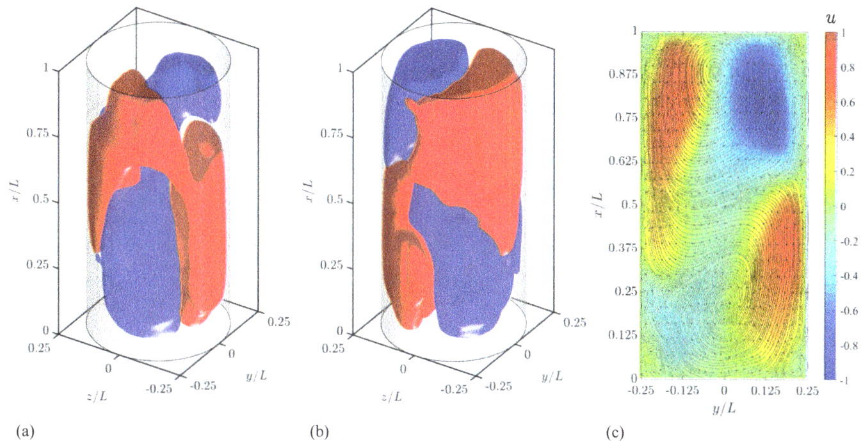

Figure 5.10: Second pair of POD modes. 3D isosurfaces of the vertical velocity component corresponding to values of $\pm 10\%$ of the maximum for (a) the 3rd and (b) the 4th mode (red corresponds to the positive value) and (c) contour map of the vertical velocity component in the $xy$-plane for the 4th mode. Values are scaled by the maximum of the vertical velocity component.

field. Interestingly, the superimposition of the first four modes can result in a torsion of the LSC as shown in figure 5.11. Here, the POD based low-order reconstructions (LORs) of a specific snapshot by using a different number of modes are reported. The LOR based only on the first two modes (figure 5.11a) is characterized by the presence of a LSC circulation laying approximately in the $xz$-plane. When the second pair of modes is added (figure 5.11b) to this field, the ascending and descending currents appear to swirl around the vertical direction; the addition of the remaining modes does not alter significantly this pattern, which in fact is present in the full snapshot (figure 5.11f).

The above considerations seem to suggest a strict relationship between the torsional oscillation of the LSC and the nature of the DRS. On a speculative level, the DRS could be regarded as the result of an extreme torsion of the LSC; this hypothesis would be corroborated by the evidence that the two rolls are not in fact separated, on the contrary, as visible in figure 5.11f, they appear twisted around each other. However, in literature the transition from the SRS to the DRS has been associated to different mechanisms. For instance, Weiss and Ahlers [195] speculated that during such a transition the single roll shrinks towards one of the plates and subsequently a new roll is established near the opposite plate; a similar but opposite evolution could explain the inverse transition from the DRS to the SRS. It is interesting to note that this mechanism in principle can be captured and reproduced by the first four POD modes. As a matter of fact, when the contributions of the two modes' pairs are such that the SR structure and a DR one lay in the same azimuthal plane, their combination causes the SR to be weakened on one side and strengthened on the opposite one. In summary, it can be argued that the first four POD modes are, in principle, capable of describing the occurrence of the LSC in the SRS, its torsional oscillation, the occurrence of the DRS (supposedly, a torsion led to the extreme point)

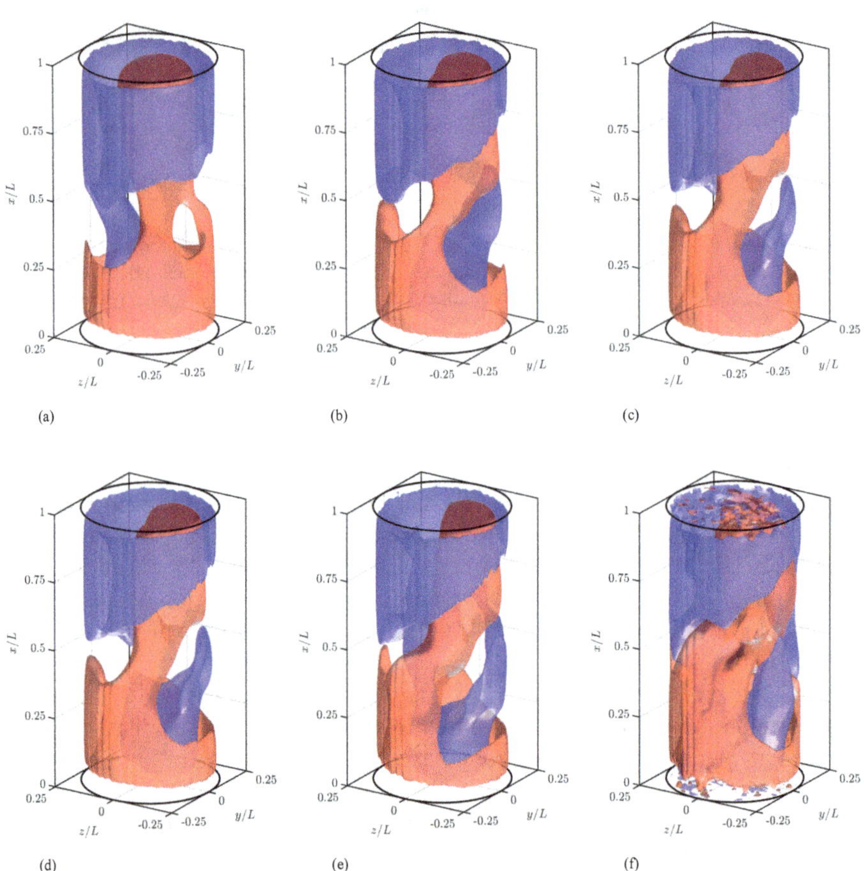

Figure 5.11: POD based low-order reconstruction of a specific snapshot by using the time-averaged velocity fields and (a) 2, (b) 4, (c) 6, (d) 8 or (e) 20 POD modes of the velocity fluctuation field and (f) instantaneous velocity field. Red and blue isosurfaces correspond to values of $\pm 0.04 u_0$.

  Chapter 5. Investigation of non-rotating Rayleigh-Bénard convection

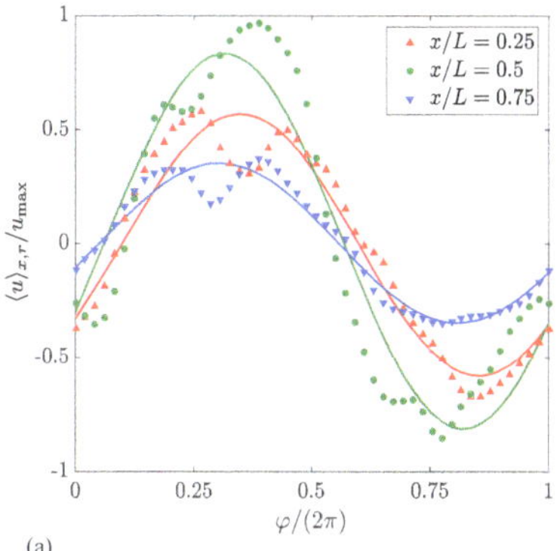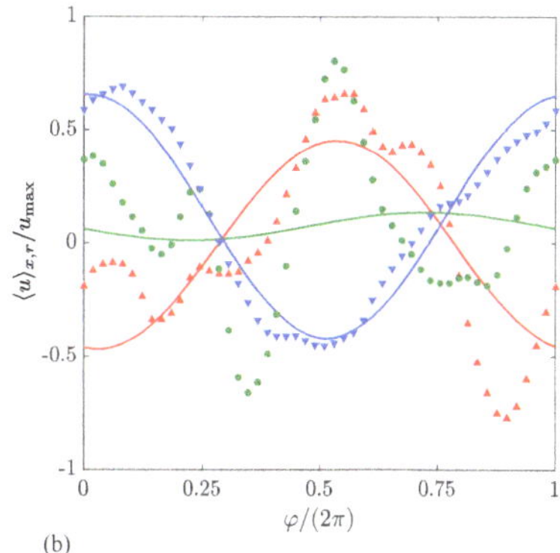

Figure 5.12: Azimuthal profiles of the vertical velocity at the heights $x/L = 0.25$ (red), $x/L = 0.5$ (green) and $x/L = 0.75$ (blue) and radial position $r/L = 0.20$ for (a) the 2nd POD mode and (b) the 4th POD mode. The data points represent the experimental measurements, whereas the solid lines are the cosine fits based on this data. Values are scaled by the maximum of the vertical velocity component in each mode.

and the transition between the SRS and the DRS.

Before moving to the analysis of the 5ht and the 6th mode, it is worth highlighting that the SR present in the first two POD modes is not a purely planar circulation, but has itself a swirled structure. This can be seen in figure 5.12a, which reports the azimuthal profiles of the vertical velocity at three different heights, namely, 0.25, 0.5 and 0.75 of the cell height, for the 2nd mode. Here, the data points indicate the experimental data (obtained by interpolation on an evenly spaced mesh), while the solid lines are the cosine fits to this data. The phase angle of these curves are clearly not coincident and the absolute difference between the phases of the signals related to the top and the bottom is about $15°$. It can be furthermore noted that the profiles are slightly different from their cosine regressions, which suggests local flow velocity variations likely due to the presence of thermal plumes. Both the above features (twisting of the SR and local flow velocity variations) are related to the fact that first POD modes, while capturing the most energetic coherent structures, are typically contaminated by smaller structures that exhibit a high degree of temporal correlation over the investigation time [197]. For the purpose of comparison, in figure 5.12b the azimuthal profiles of the vertical velocity at the same heights and radial position of figure 5.12a are reported for the 4th mode. In such a case, the cosine fits of the profiles related to the top and the bottom are shifted by nearly $180°$, whereas the profile at $x/L = 0.5$ has a distribution very different from its cosine fit, since it corresponds to the flow region between the two rolls.

Figure 5.13 shows the 5th and the 6th POD mode. These two modes are characterized by the presence of four adjacent vertical currents localized in separate angular sectors of the cylindrical sample, as clearly visible in the section of the velocity field at mid-height (Figure 5.13c). The vertical currents generate four large-scale rolls, two by two with opposite circulation, which are distributed azimuthally around the cylinder axis. Interestingly, these rolls, when superimposed to the flow pattern given

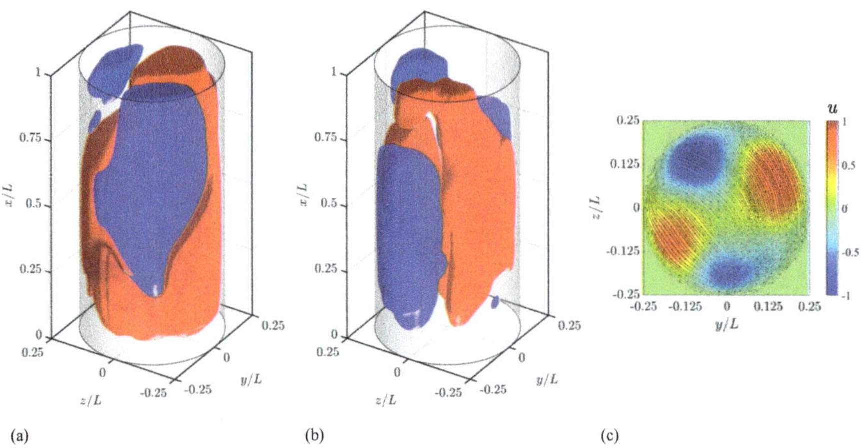



Figure 5.13: Third pair of POD modes. 3D isosurfaces of the vertical velocity component corresponding to values of $\pm 10\%$ of the maximum for (a) the 5th and (b) the 6th mode (red corresponds to the positive value) and (c) contour map of the vertical velocity component in the $yz$-plane for the 6th mode. Values are scaled by the maximum of the vertical velocity component.

by the first four modes, can be responsible of an out-of-plane motion of the LSC, which squeezes it onto a side of the cylindrical cell.

To show this, LORs of a specific snapshot by varying the number of modes employed are reported in figure 5.14. In this case, it can be noted how the addition of the contribution due to the second pair of POD modes (compare figures 5.14a and 5.14b) does not change significantly the structure of the LSC, which lays approximately in the azimuthal plane $y = -z$. However, when the contributions related to the 5th and the 6th mode are added (figure 5.14c), the LSC is observed to move in the direction normal to its plane and approach the sidewall on one side. The addition of two more POD modes (figure 5.14d) does not introduce significant changes, whereas the high-order POD modes (figure 5.14e-f) appear to modify the region of the flow opposite to the squeezed LSC. The case reported in figure 5.14 shows that the 5th and the 6th mode are responsible of the sloshing mode of the LSC, that have been reported in numerous works ([52, 53, 198]).

In conclusion, it should be mentioned that the pairing of two POD modes is often investigated by analyzing the trajectories in the phase planes identified by the corresponding POD temporal coefficients (see the appendix A for a definition of POD coefficients and more details). Often, such an analysis leads to the identification of Lissajous curves in the phase planes, which suggest that the pairs of POD modes exhibit complex harmonic evolution and typically allow to associate each mode with a specific frequency. In the present case, when trajectories on the phase planes are represented, no coherent patterns are indeed detected (corresponding figures are not reported here for brevity); conversely, a quite random distribution of the states in the phase plane is observed for any pair of POD modes examined. This is due to the fact that turbulent thermal convection in a confined geometry is a chaotic flow, which does not exhibit predominant modes occurring with constant oscillation frequencies;

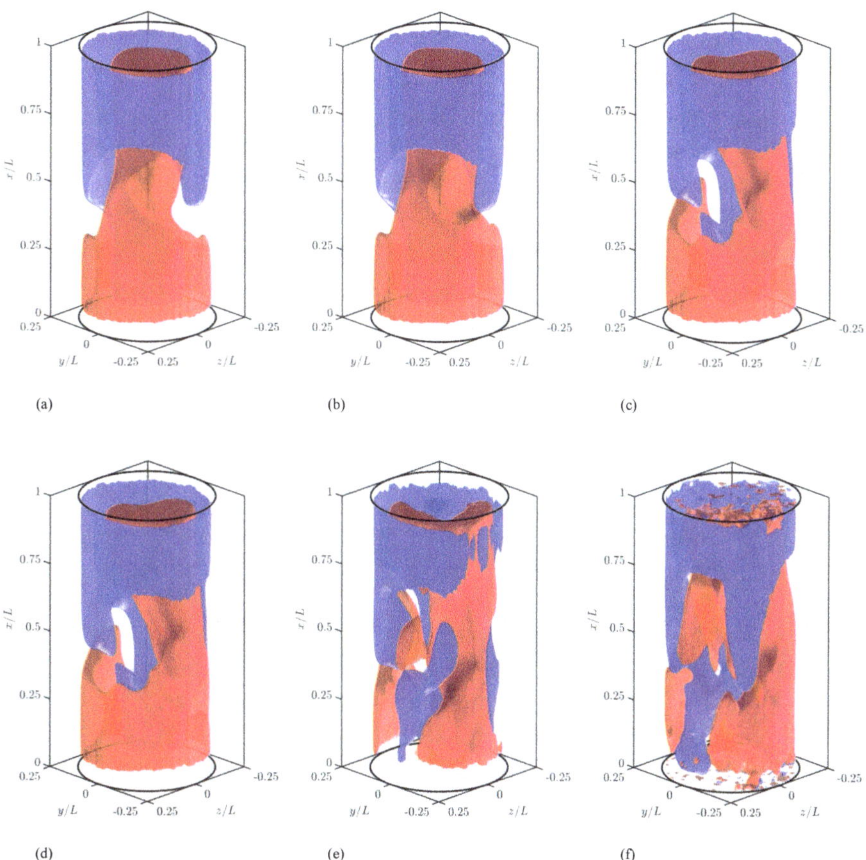

Figure 5.14: POD based low-order reconstruction of a specific snapshot by using the time-averaged velocity fields and (a) 2, (b) 4, (c) 6, (d) 8 or (e) 20 POD modes of the velocity fluctuation field and (f) instantaneous velocity field. Red and blue isosurfaces correspond to a values of $\pm 0.04 u_0$. This case shows how, for the selected snapshot, the addition of the contribution of the 5th and 6th modes causes an out-of-plane displacement of the LSC due to the mean and the first four modes' contributions.

rather, it is characterized by multiple states alternating in a very complex way. Insight into such a chaotic evolution is indeed possible only through a statistical analysis, as done in the following.

## 5.3. Statistical behavior of the LSC

$\mathrm{T}$HIS section focuses on the oscillating behavior of the LSC. In experimental investigations, the LSC orientation and oscillations are identified by measuring the azimuthal temperature distribution at the sidewall [46, 51, 93]. Thermal probes are typically arranged with uniform angular spacing along circumferences at three different heights, namely $x = 0.25\,L$, $x = 0.5\,L$ and $x = 0.75\,L$; the most common number of probes is 8. The discrete temperature measurements $\theta^{(k)}$ are then fitted to a cosine function:

$$\theta^{(k)} = \theta_j + A_{\theta j} \cos(\varphi^{(k)} - \varphi_{\theta j}) \tag{5.1}$$

where $\varphi^{(k)}$ are the angular positions of the probes and $\theta_j$, $A_{\theta j}$ and $\varphi_{\theta j}$ are the average temperature, the amplitude and the initial phase of the cosine fit at the $j$-th height where measurements are taken. In the following, the indices $b$, $m$ and $t$ are used in place of the generic index $j$ to denote the above quantities in correspondence of the distribution at $x = 0.25\,L$ (bottom), $x = 0.5\,L$ (middle) and $x = 0.75\,L$ (top). The LSC orientation is identified by the angle $\varphi_{\theta m}$, while the differences between each pair of the angles $\varphi_{\theta b}$, $\varphi_{\theta m}$ and $\varphi_{\theta t}$ provide insight in the oscillations of the LSC.

In our investigation, temperature measurements are not available, since the requirement of optical accesibility makes it impossible to place thermal probes in the cylinder sidewall. However, velocity measurements are carried out within the entire cell volume; therefore, virtual anemometric probes can be placed at the same heights and at a selected distance from the cylinder axis to sample the vertical velocity component at each measurement time. A fitting procedure similar to that expressed by equation (5.1) can be applied to the azimuthal vertical velocity profile, thus obtaining a velocity amplitude $A_{u j}$ and an azimuthal orientation of $\varphi_{u j}$ of the LSC at the $j$-th height. In their numerical investigation, Stevens *et al.* [199] demonstrated that the two approaches lead essentially to the same results, at least from a statistical point of view.

Since the present investigation is carried out in a relatively low $Ra$ regime, as suggested by Stevens *et al.* [199] it is necessary to use a number of probes significantly larger than 8 to have sufficient resolution to distinguish between the SRS and the DRS of the LSC. In the present case, 98 probes are indeed employed, which corresponds to an angular spacing of about $3.6°$ and linear spacing of about $2.13$ mm (30 voxels) on the selected circumference (the radius of which is equal to $0.9$ times the internal radius of the cylinder). The point measurements are obtained by interpolating the Lagrangian trajectories as explained in section 3.1.4, thus the

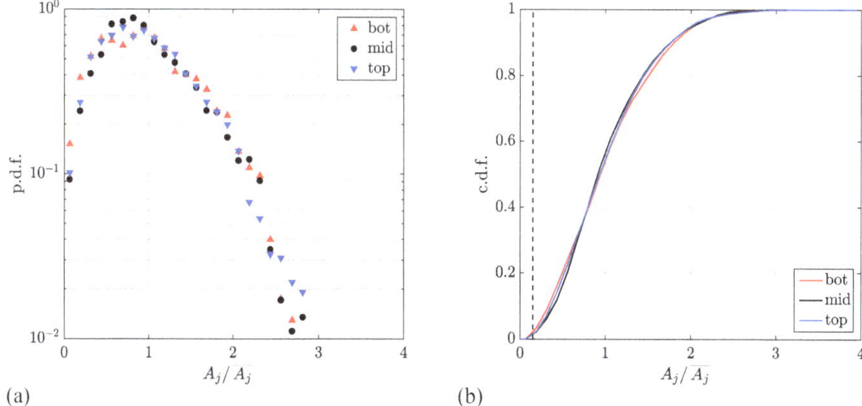

Figure 5.15: Probability distribution of the vertical velocity amplitudes of the LSC at the bottom ($x/L = 0.25$), middle ($x/L = 0.5$) and top ($x/L = 0.75$): (a) probability density function; (b) cumulative density function. Each amplitude is normalized by the corresponding time-averaged value.

azimuthal profiles can be very noisy in the instantaneous velocity field. In order to reduce noise a moving time-averaging filter is used as still suggested in [199].

In addition to the time-resolved behavior of the quantities $A_{uj}$ and $\varphi_{uj}$ (in the following also denoted simply by $A_j$ and $\varphi_j$), the time-resolved relative LSC strength $S_j$ is calculated again as proposed by Stevens *et al.* [199]. $S_j$ is the ratio of the energy in the first Fourier mode of the azimuthal vertical velocity profile to the the total energy in all Fourier modes and a value of $S_j$ greater than 0.5 indicates a high goodness of the cosine fit (condition necessary to the occurrence of the LSC).

Of course, the strength of the LSC is also related the vertical velocity amplitudes $A_j$. According to Weiss and Ahlers [63], the existence of the LSC in the SRS or the DRS implies that all the three amplitudes $A_j$ for the top, the middle and the bottom have to be higher than a certain threshold, otherwise a transitional state of the turbulent convection is observed. Weiss and Ahlers [63] fixed as reasonable threshold 15% of the time-averaged amplitude $\overline{A_j}$ and they defined as "events" all the realizations in which one of the three amplitudes is lower than such a threshold. In particular, the simultaneous drop of all the amplitudes below the threshold is associated with a cessation of the LSC.

## 5.3.1. Statistical behavior of the LSC strength

$\mathbf{F}$IGURE 5.15 reports the probability and the cumulative distribution functions of the amplitudes $A_j$ for the top, middle and the bottom. With the chosen scaling ($A_j$ normalized by its time-averaged value $\overline{A_j}$), the three probability distributions essentially collapse on each other, as already observed in the work of Weiss and Ahlers [63] for $Ra = 9 \times 10^{10}$, $Pr = 4.38$ and $\Gamma = 0.5$ (figure 10 of their article). Differently from the latter case, however, the maximum of the p.d.f is reached at a value smaller than 1. In figure 5.15b, where the cumulative density functions

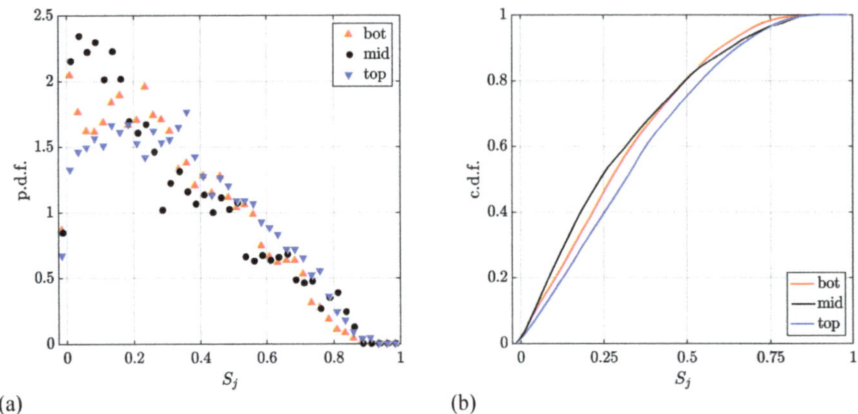

Figure 5.16: Probability distribution of the LSC strength at the bottom ($x/L = 0.25$), middle ($x/L = 0.5$) and top ($x/L = 0.75$) of the cell: (a) probability density function; (b) cumulative density function.

are reported, the dashed line represents the threshold chosen by Weiss and Ahlers [63] to identify an event. The probability that each amplitude is lower than the threshold is $3.15\%$, $1.73\%$ and $2.98\%$ for the bottom, middle and top, respectively. These values correspond to frequencies of such events that, non-dimensionalized by the viscous time $\tau_\nu = L/\nu$, are equal to 112, 53 and 106. Those values are very similar to that observed by Weiss and Ahlers [63] and Xi and Xia [58] in their works at higher Rayleigh numbers ($> 10^9$) and lower Prandtl number ($\approx 4.3$) within a cylinder with aspect ratio equal to one-half. The only difference is that the occurrence frequency of the events at mid-height is found to be remarkably lower than (half) those at the bottom and the top. Moreover, no cessations (i.e., events when all the three amplitudes are reduced below their thresholds) are observed in the present investigation. This seems to be in agreement with the experimental observations of Weiss and Ahlers [63] and Xi and Xia [58], who found a very small frequency of the cessations. It should be remarked that the duration of the T-PIV experiment is lower than that of the investigations cited above; therefore, the reported values of the frequencies of the events represent an indication rather than a robust statistics.

The statistical behavior of the LSC strengths $S_j$ can be inferred from figure 5.16. The behavior of $S_m$ (black data points) looks very similar to that reported in [199] for a numerical investigation carried out in analogous operating conditions (but with adiabatic zero-thickness wall). $S_m$ can be considered as representative of the strength of the LSC in the SRS, while $S_t$ and $S_b$ are related to both the SRS and the DRS. At a closer look to the figure 5.16, it is indeed possible to note that the p.d.f of $S_m$ exhibits higher values for $S < 0.2$; this is essentially related to the occurrence of the DRS which causes $S_m$ to have values smaller than those of $S_t$ and $S_b$.

Focusing on the c.d.f.s (figure 5.16b), $S_m$ and $S_b$ appear to have very similar probability distributions for $S_j > 0.5$ (i.e., in the region of a high goodness of the cosine fits), while $S_t$ exhibits smaller values of the cumulative probability density.

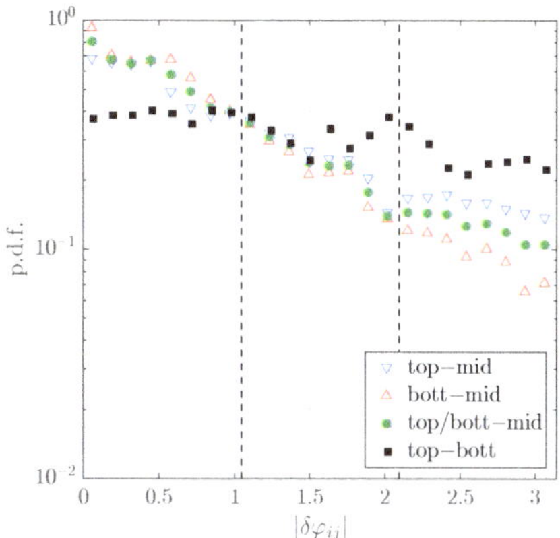



Figure 5.17: Probability distributions for the absolute difference between the orientation of the LSC at the bottom ($x/L = 0.25$), middle ($x/L = 0.5$) and top ($x/L = 0.75$). Only the samples satisfying the criterion for the amplitudes (see the text for explanation) have been considered. When exceeding $\pi$, the difference $|\delta\varphi_{ij}|$ was replaced by $2\pi - |\delta\varphi_{ij}|$. The dashed lines indicate the definition of the SRS ($|\delta\varphi_{ij}| < \pi/3$ for any $i \neq j$) and the DRS ($|\delta\varphi_{tb}| > 2\pi/3$) according to [58] and [63]. Phase differences are in radians.

In particular, for any value $\overline{S}$ such that $0.5 < \overline{S} < 0.75$, the probability that $S_t > \overline{S}$ is greater than the probabilities that $S_m > \overline{S}$ and $S_b > \overline{S}$ (this indeed holds also for $\overline{S} < 0.5$). At a first glance, such a statistical behavior might be related to the influence of the sidewall on the flow field. In section 5.1, it has been shown that the condition of a sidewall external temperature smaller than the average temperature introduces an asymmetry of the flow field between the top and the bottom half of the cell, since the cold plumes are subjected to a reduced heat transfer with the sidewall than the hot plumes. As a consequence, the strength of the LSC might be expected statistically higher in the top half of the cell than in the bottom half. However, such an argument is not supported by the numerical simulations carried out at the same conditions of this experiment, as shown in the following (see section 5.4).

## 5.3.2. Statistical occurrence and properties of the SRS and DRS

IN order to identify the occurrence of the SRS and the DRS in the instantaneous evolution, the criteria introduced by Xi and Xia [58] and Weiss and Ahlers [63] are used here. Therefore, the LSC is recognized to be in the SRS when the three amplitude $A_j$ are simultaneously above the thresholds $0.15\,\overline{A_j}$ and the differences $|\delta\varphi_{ij}| = |\varphi_i - \varphi_j|$ for each $i \neq j$ are lower than $60°$. Conversely, the occurrence of the DRS requires $|\delta\varphi_{tb}| > 120°$ and the amplitudes of the top and the bottom vertical velocity profiles to be above their thresholds.

After the above definitions, the p.d.f.s of the absolute differences $|\delta\varphi_{ij}|$ provide insight into the statistical occurrence of these two modes of the LSC. Consequently, such distributions have been reported in figure 5.17. More specifically, in order

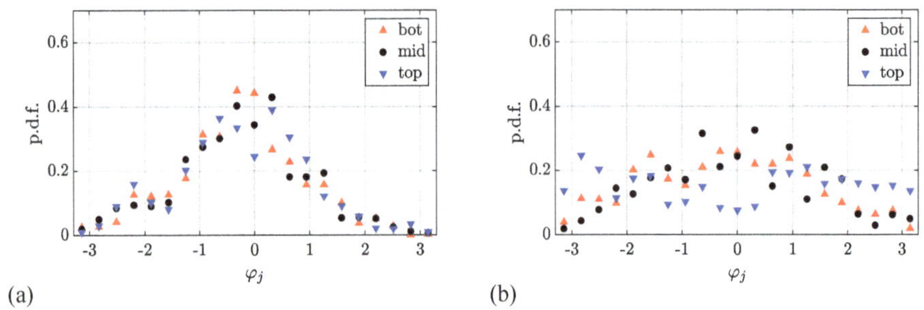


Figure 5.18: Probability distributions for the LSC orientation at the bottom ($x/L = 0.25$), middle ($x/L = 0.5$) and top ($x/L = 0.75$) for (a) the SRS and (b) the DRS. Angles are in radians.

to construct such a diagram only the samples satisfying the criterion for the LSC amplitude have been considered. The distributions of $|\delta\varphi_{tm}|$ and $|\delta\varphi_{bm}|$ are practically coincident and they exhibit a broad Gaussian-like shape. This is in very good agreement with the experimental results of Weiss and Ahlers [63] (figure 21) and the numerical results of Stevens *et al.* [101] (figure 11). On the other side, $|\delta\varphi_{tb}|$ shows a nearly uniform distribution with a mild decrease towards high values of the phase difference. As a consequence of the latter evidence, based on the criterion above introduced, the probability of occurrence of the DRS (i.e., the probability of the event $|\delta\varphi_{tb}| > 120°$) is expected to be slightly lower than 30%. Indeed, the computed value is 29.3%, whereas the relative frequency of occurrence of the SRS is 27.1%. This means that the transitional states have a probability of occurrence of 43.7% among states that are not classified as events. Such results are again in agreement with the much longer experiments of Weiss and Ahlers [63], who found that for $Ra = 1 \times 10^8$ and $Pr = 4.38$, the flow state is undefined for about the 50% of the time (in this regards, it is remarked that the events cover only the 6% of the total time in the present experiments, thus the transitional states constitute the 41% of the time).

In figure 5.18 the probability distributions of the LSC orientations at the top, middle and bottom are presented for both the SRS and the DRS. As expected, the distributions for the SRS (figure 5.18a) are essentially the same at all the three levels and they exhibit a pronounced peak around $\varphi = 0°$, while the p.d.f.s go to zero towards $\pm180°$. It is difficult to argue why $\varphi = 0°$ is a preferential orientation in the present case. Brown and Ahlers [46] demonstrated that symmetry-breaking inhomogeneities can be generated by several factors, among which are the Earth's Coriolis force and imperfections of the experimental apparatus like a slight tilt of the sample, small horizontal thermal gradients in the top and the bottom plate and eccentricity of the circular cross-section. The investigations of Xi and Xia [58] and Weiss and Ahlers [63] have shown that the first effect (Earth's Coriolis force) seems to be smaller in the case of $\Gamma = 0.5$ than for $\Gamma = 1$. Among the small imperfections of the experimental apparatus, the existence of small horizontal thermal gradient in the top plate cannot be ruled out, although the structure of the heat exchanger would get one to expect radial inhomogeneities rather than azimuthal ones. As

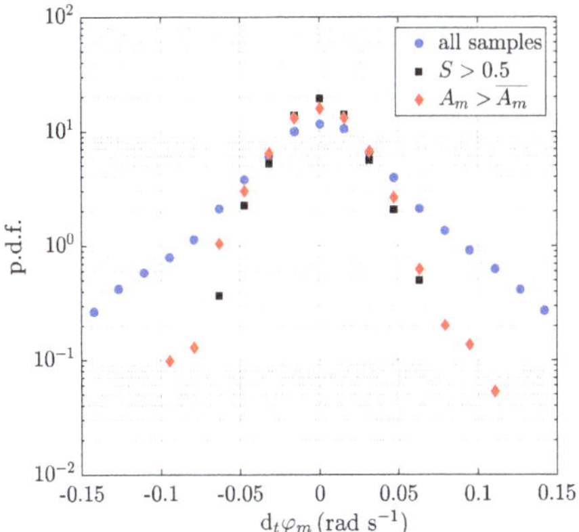

Figure 5.19: Probability distribution for the angular velocity of the LSC.

regards the tilt angle, the cell was leveled with an accuracy of $0.1°$, while no data is available on the non circularity of the cross section from the producer. Ultimately, we cannot rule out that, in the present case, Earth's rotation has no influence on the LSC orientation. Interestingly, the p.d.f. of the LSC orientation for the DRS (figure 5.18b) is remarkably uniform for all the three levels. Although, at a closer look, a mild dip for the top distribution and a rise for the bottom and middle ones are observed around $\varphi = 0°$, it is possible to conclude that it does not exist a preferential orientation for the DRS.

Figure 5.19 reports the p.d.f. of the LSC angular velocity, estimated as the time derivative of the LSC orientation at mid-height. The distribution has a Gaussian shape in the central part with fairly heavy tails, in agreement with the results from the previous investigations of Xi and Xia [58] and Weiss and Ahlers [63]. In the same diagram, the p.d.f. for the samples satisfying the Stevens $et\ al.$'s criterion for identification of a coherent LSC based on circulation strength, i.e. $S_m > 0.5$, (black squares) and that related to the samples with $A_m > \overline{A_m}$ (red diamonds) are also reported. Such curves exhibit an almost identical narrower Gaussian shape, which indicates that the high values of the angular velocity of the LSC indeed are related to states of the turbulent convection where a well defined LSC is indeed not found and thus the determination of the LSC orientation is affected by greater uncertainty.

### 5.3.3. Relationship of the SRS and the DRS with low order POD modes

IN section 5.2, it has been shown that the first POD modes are essentially related to the SRS and the DRS of the LSC. In the preceding paragraphs of the present section, the identification of the SRS and the DRS have been performed via the criteria adopted in previous works on this subject. It is interesting to analyze the relationship of the thus identified SRSs and DRSs with the first two pairs of POD modes. For this purpose, we can focus on the statistical behavior of the corresponding

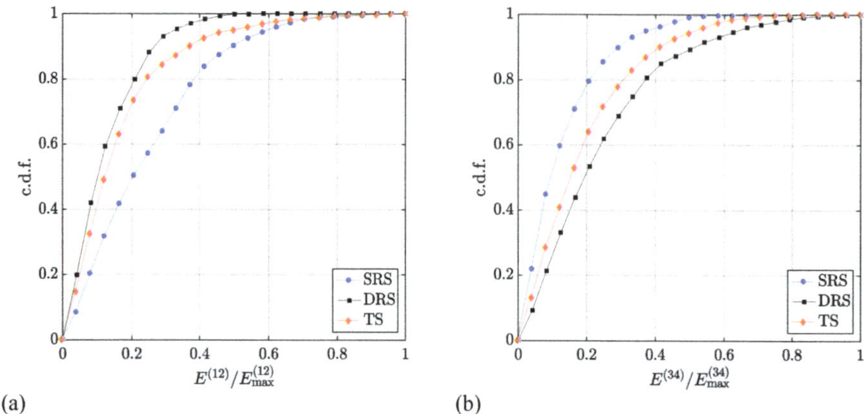

(a)          (b)

Figure 5.20: Probability distributions of the energy contribution due to the (a) the first pair and (b) the second pair of POD modes for the different states of the LSC. Energies are scaled by the maximum value detected over the entire measurement duration.

POD coefficients, which are the projection of the snapshots onto the POD modes (see appendix A for more information). Since the POD represents an orthogonal decomposition, such coefficients are representative of the energy contribution of each mode to the total kinetic energy of the instantaneous flow field. Following the observation that the first POD modes appear to be paired, it is reasonable to define an energy contribution related to the pairs of the POD modes and study the statistical distribution of this quantity. With the notation adopted below, $E^{(ij)}$ represents the energy contribution due to the pair of the $i$-th and the $j$-th POD mode.

Figure 5.20 reports the c.d.f. of the energy contributions $E^{(12)}$ and $E^{(34)}$ distinguishing between the samples characterized by the SRS, the samples characterized by the DRS and those related to a transitional state (TS), i.e. a state that is not an event nor the SRS or the DRS (according to the definitions introduced in the previous paragraph). It is here remarked that a high value of the c.d.f. corresponding to a certain energy level $\overline{E}$ means that the probability that $E^{(ij)} > \overline{E}$ is small (this probability is equal to $1 - \mathrm{c.d.f.}(\overline{E})$). Consequently, figure 5.20a shows that the first pair of POD modes statistically contributes more to the SRS than to the DRS. On the other side, figure 5.20b shows that the second pair of POD modes statistically contributes more to the DRS than to the SRS. This is consistent with the observation that a high contribution from the second pair of POD modes can cause either a considerable torsion of the LSC to the point of setting it in the DRS or a shrinking of the single roll structure towards one of the two plates according to the mechanism sketched by Weiss and Ahlers [63] (figure 9 of their article). Another interesting point emerging from figure 5.20 is that the energetic contributions of the POD modes' pairs to the TSs are on an intermediate level between those related to the SRS and the DRS in a statistical sense. The above considerations are essentially a confirmation that the most energetic POD modes capture the main states of the turbulent convection.

Another useful perspective on the relationship between the states of the LSC

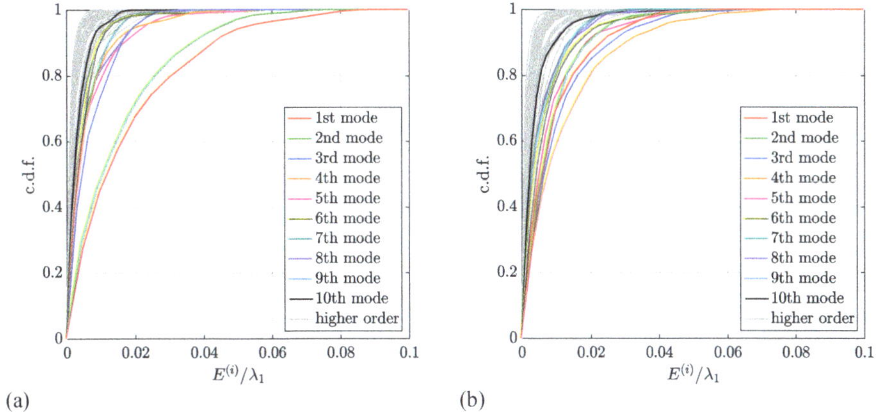

Figure 5.21: Probability distributions of the mode energies for (a) the SRS and (b) the DRS. Energies are scaled by the first POD eigenvalue.

and the POD modes is obtained by representing the c.d.f. of the different mode energies $E^{(i)}$ for all the samples corresponding to the same state of the LSC. Such a representation is given in figure 5.21 for the SRS and the DRS; here, energy values are scaled by the first POD eigenvalue, which represents the total mean energy contained in the first POD mode. Figure 5.21a shows that the contribution of the first pair of POD modes is statistically preponderant with respect to the contributions of the remaining modes. This could seem an obvious consideration since the POD modes retain most of the time-averaged fluctuating kinetic energy. However, it should be remarked that the c.d.f.s in figure 5.21a were constructed by using only the samples in a SRS and *not* the entire collection of samples available. When a subset of samples is indeed considered, it is not true in general that the first two POD modes contain the greatest amount of kinetic energy. On the other hand, this is proved by figure 5.21b where indeed it is noted that the major contribution from the energetic point of view to the subset of samples corresponding to the DRS is given by the 3rd and the 4th mode. The observation that the first two POD modes are strictly related to the SRS leads to the conception of a different way of detecting the LSC orientation, which is based on the correlation of the snapshot with the first two POD modes. Such a technique and its relationship with the method employed in section 5.3.2 is discussed in the next paragraph.

## 5.3.4. Identification of LSC orientation based on POD

An alternative method of detecting the LSC orientation could be founded on LOR of snapshots based on only the first two POD modes, which, as shown in the previous paragraph, capture most of the fluctuating kinetic energy in the SRS. Theoretically, each snapshot can be reconstructed by the mean and the first two POD modes, subsequently the cosine fit of the azimuthal vertical velocity profile at the middle level provides an LSC orientation that is filtered out of the contributions of

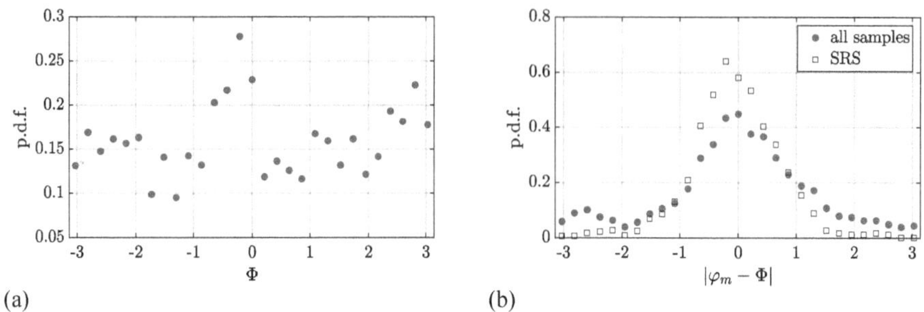

Figure 5.22: Statistical behavior of the LSC orientation identified by means of the first two POD modes: (a) probability density function; (b) comparison with the LSC orientation identified by the cosine fit of the azimuthal velocity profile at $x/L = 0.5$. Angles are in radians.

the oscillating modes (torsion and sloshing), which are represented by the high order POD modes as previously explained. However, in practical implementation, it is possible to combine simply the azimuthal profiles of the axially and radially-averaged vertical velocity for the 1st and the 2nd mode, i.e. the profiles reported in figure 5.9. More specifically, these profiles are summed by weighting them through the corresponding mode coefficients for the selected snapshot; then the LSC orientation is obtained as the phase angle corresponding to the maximum in the resulting profile. It should be commented that the same procedure can be performed considering the azimuthal profiles of the vertical velocity at mid-height and a specific radial position. Nevertheless, due to the residual swirling component of the single roll structure in the first pair of POD modes (see figure 5.12a), we believe that it is more robust to employ the radially- and axially-averaged profiles.

Figure 5.22a presents the probability distribution of the LSC orientation based on the method described above (denoted by the symbol $\Phi$). It can be noted that, differently from the distribution of $\varphi_m$ (figure 5.18a), that of $\Phi$ does not feature a well-defined Gaussian distribution; this is also due to the fact that all the samples were employed to construct such a p.d.f.. The absence of a clear behavior of the p.d.f. suggests that also the method based on POD fails in detecting a correct LSC orientation in the transitional states. However, a preferred orientation is still detected, approximately at $\Phi \approx -12°$. This is indeed corresponds to the phase angle of the blue curve in figure 5.9. Such a result is not surprising if we consider that when a preferential orientation exists the first POD mode is prone to capturing such an orientation. The fact that the method based on the cosine fit of the azimuthal velocity profile at mid-height does not present a prominent peak at the LSC orientation of the first mode could indicate a bias error in the estimation of the LSC orientation. However, it is very difficult to state if this is just a conjecture or not. Indeed, when the p.d.f. of the difference between $\Phi$ and $\varphi_m$ is plotted (figure 5.21b, blue dots), a peak at the above-mentioned angle is not observed. Nevertheless, by restricting to only the samples related to the SRS, a peak is instead found. The origin of such a bias is not clear, but a plausible explanation could be that the level $x/L = 0.5$ in the present case is not suitable for the identification of the LSC orientation because of

    CHAPTER 5. Investigation of non-rotating Rayleigh-Bénard convection the axial asymmetry introduced by the specific thermal condition of the sidewall.

## 5.4. Comparison with numerical simulations

IN this section, the experimental measurements presented above are compared with the results of direct numerical simulations performed in similar conditions. These simulations include the presence of a sidewall with physical properties equal to that of the cylindrical sample used in the present setup. The boundary condition imposed at the external side of the sidewall is uniform temperature equal to the temperature of the tank measured during the experiment. In non-dimensional terms, this corresponds to $\theta(\varphi, r = r_i + \Delta r, x) = \theta_e = 0.48 \quad \forall \varphi, \forall x$. Complete information about the boundary conditions used in the simulation is reported in Table 3.2. It is important to underline that, while the physical behavior of the sidewall is correctly modeled in the simulation, the imposed thermal boundary condition could be different from the actual one, since the presence of thermal gradients on the external surface of the cylinder cannot be ruled out. On the other hand, mapping the external temperature distribution of the cylinder sidewall (with an appropriate number of thermal probes) is unfeasible due to the requirement of optical accessibility.

In section 5.1, a good agreement between the mean fields of the vertical velocity component and its fluctuation determined experimentally and numerically has already been shown. In the following, the POD analysis and the study of the statistical behavior of the LSC are repeated for the numerical simulations and differences with the experimental observations are appropriately highlighted.

### 5.4.1. Extended POD modes

SINCE in numerical simulations temperature is determined in any point of the flow field (and also inside the physical sidewall), the correlation of the temperature fluctuations with the velocity fluctuations can be studied by means of the extended POD modes [200–202] (the reader is referred to appendix A for further information). Figure 5.23 presents the velocity POD modes obtained from the numerical simulation. The flow structure in each mode is still represented by 3D isosurfaces of positive and negative vertical velocity, which identify the ascending and descending currents in the modal field; however, such isosurfaces are now flooded by the values of the corresponding extended POD modes of the temperature fluctuation. These have been determined essentially by projecting the POD "temporal" basis of the velocity fields onto the instantaneous temperature fields. Figure 5.23 essentially confirms that the ascending currents carry fluid warmer than the bulk, while the descending ones consists of colder fluid. However, the extended POD modes (here non represented for brevity) exhibit a distribution significantly less smooth than that of the velocity modes, because of the local variations of the temperature field related to the plumes that are clustered in or carried by the vertical currents.

Interestingly, the first two POD modes from the numerical simulation (figure 5.23a-b) have the same morphology of those obtained by the experimental measure-

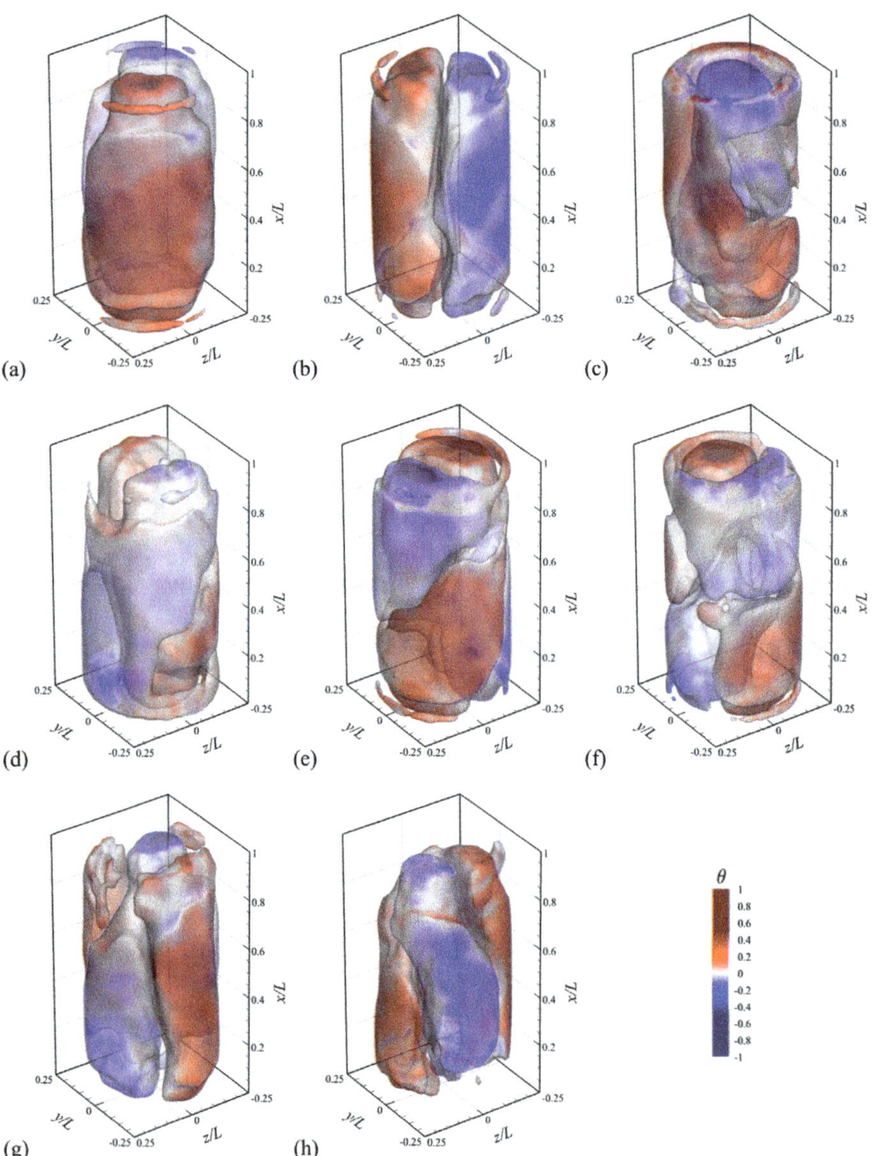

Figure 5.23: Velocity POD modes from the numerical simulation. 3D isosurfaces of the vertical velocity component corresponding to $\pm 7\%$ of the maximum for the first eight modes (from (a) to (h), respectively). Colors correspond to values of the extended POD mode of the temperature fluctuations, normalized by the maximum for each mode.

ments, with a SR structure filling the entire domain. The only evident difference is a rotation of $90°$ about the cylinder axis. Conversely, the modes 3-4 have a quite different structure from the experimental counterparts. In the 3rd and the 4th POD mode (figure 5.23c-d) the DRS is now replaced by twisted currents that resemble the torsional mode of the LSC. A DRS is remarkably noticed in the two successive modes (figure 5.23e-f). As noted in section 5.2, also these modes are able to describe a torsional oscillation of the LSC, we cannot exclusively associate the 3rd and the 4th numerical POD mode with such an oscillatory mode. However, since the energy of POD modes decrease with increasing order, the pattern related to the 3rd and the 4th mode has a higher degree of correlation with the instantaneous velocity fields than that related to the 5th and the 6th mode (i.e., to the DRS). The 7th and the 8th mode (figure 5.23e-f) are very similar to the 5th and the 6th experimental POD mode and thus they are related to the off-plane displacement of the LSC.

We may summarize the above results by concluding that the high order POD modes from the numerical simulations are essentially the same of the experimental case, although an additional pair of POD modes indicating a torsion of the LSC is found in the former case. The reason of such a discrepancy is not simple to predict, but however it does not represent a major difference between the experimental and the numerical observations.

Another interesting perspective on the characteristic modes of the thermal convection is offered by the analysis of the POD modes of the temperature fluctuations and the corresponding extended velocity POD mode. Such modes are represented in figures 5.24 and 5.25. The first two POD temperature modes do not feature an LSC as main structure. Conversely, in these modes it is possible to detect some structures that arise in the center of the top and the bottom plates or the corner between these plates and the sidewall and extend over the entire volume of the cell in an intricate path. Such structures are still ascribable to the plumes forming at the thermal boundary layers on the plates, which, under the effect of their buoyancy forces, propagate from one plate to the other. In particular, in the first extended mode the structures arising from the center of one plate reach the opposite plate in correspondence of the corner and vice versa. It is noted that, even though the field represented in figure 5.24a-b is characterized by positive temperature fluctuations in the center of the bottom plate and negative temperature fluctuations in the outer region adjacent to the corner, the contribution given by such a mode to a generic snapshot can be opposite in sign since the mode is weighted by a coefficient that occurs to be negative at certain time instants. Therefore, generally speaking, it can be concluded that, in this characteristic mode, heat is convected from the central part of the bottom plate to the corner of the top one and from the corner of the bottom plate to the central part of the top one. On the contrary, in the second mode, heat is convected from the center of the bottom plate to the center of the top one and from corner to corner. It is then evident that the first two POD temperature modes are not paired differently from the first POD velocity modes; on the contrary, they capture different mechanisms of the convective heat transfer relying on a different structure of the velocity field.

Focusing on the modes 3-6 (figure 5.25), it is possible again to detect the SRS

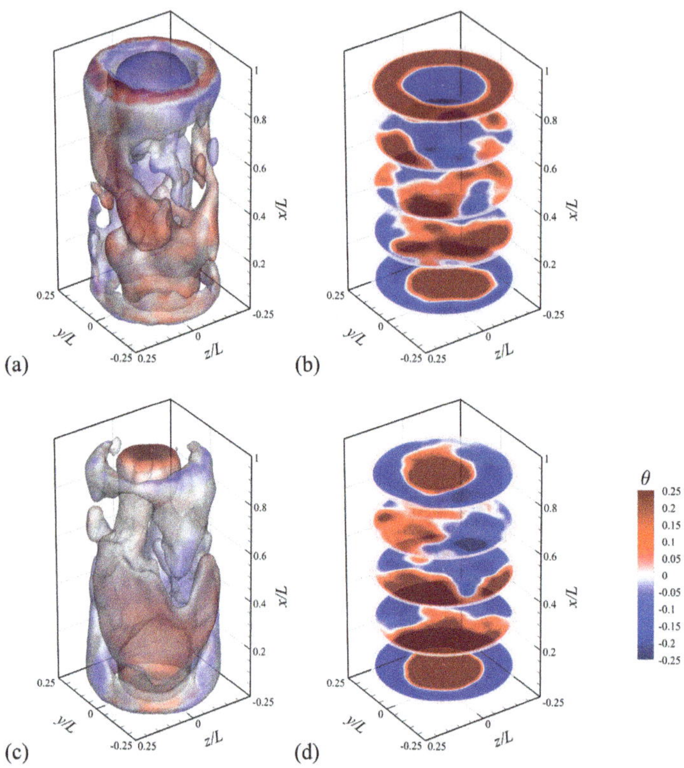

Figure 5.24: Temperature and extended velocity POD modes from the numerical simulation. 3D isosurfaces of the vertical velocity component for (a) the 1st and (c) the 2nd mode and slices of the temperature field for (b) the 1st and (d) the 2nd mode. Colors correspond to values of the temperature fluctuations, normalized by the maximum for each mode.

and the DRS of the LSC. Interestingly, the 5th and the 6th temperature mode exhibit a well-defined DRS, similarly to the 3rd and the 4th experimental velocity modes.

It is worth noting that the presence of the same flow structures in both the extended POD temperature and velocity modes indicate that these structures contribute fundamentally not only to the temporal fluctuation of the temperature and the velocity but also to their correlation, i.e. to the convective heat transfer. However, the absence of a LSC in the first two POD temperature modes means that the main contribution to the temperature fluctuation is not given by the LSC itself. Indeed, the flow morphology detected in these modes could be associated with the transitions between the different states (SRS and DRS) of the LSC, suggesting that the highest temperature fluctuations in the flow field occur in correspondence of these states.

## 5.4.2. Statistical behavior of the LSC

THE analyses carried out in section 5.3 have been repeated for the numerical simulations. In general, a very good agreement with the experimental measurements

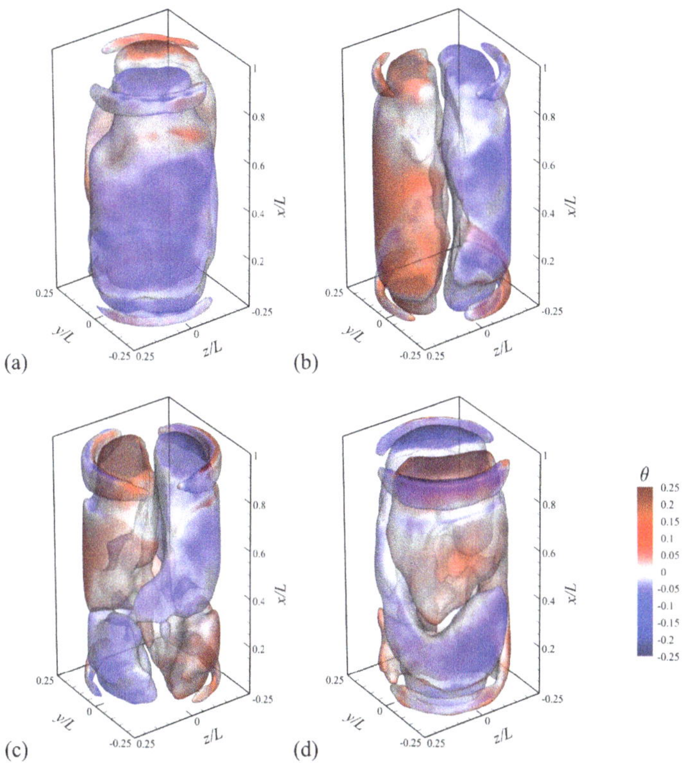

Figure 5.25: Temperature and extended velocity POD modes from the numerical simulation. 3D isosurfaces of the vertical velocity component for the modes 3-6 (from (a) to (d), respectively). Colors correspond to values of the temperature fluctuations, normalized by the maximum for each mode.

has been found. In the following, focus is given essentially on the discrepancies observed between the two sets of data.

A first remarkable difference concerns the statistical distribution of the LSC strengths at the three levels $x/L = 0.25$, $x/L = 0.5$ and $x/L = 0.75$. The p.d.f.s and the c.d.f.s of these strengths for the numerical simulation are reported in figure 5.26. Differently from the experimental measurements, the three curves show a very similar trend, except for the range $S_j < 0.1$ where the p.d.f. related to the middle level reaches higher values than those corresponding to the top and the bottom. As explained before, this is ascribed to the occurrence of the DRS. On the other hand, the c.d.f.s for the top and the bottom are found to be essentially coincident with each other. Since the present simulation includes the effects of the physical sidewall, this results suggest that the statistically higher values of the top LSC strength found in the experimental case (section 5.3.1) are not due to the influence of the cylinder wall thermal condition on the flow field. Indeed, they are likely to be caused by other inhomogenities or asymmetries of the experimental apparatus. Among the most plausible explanations are the presence of horizontal thermal gradients in the

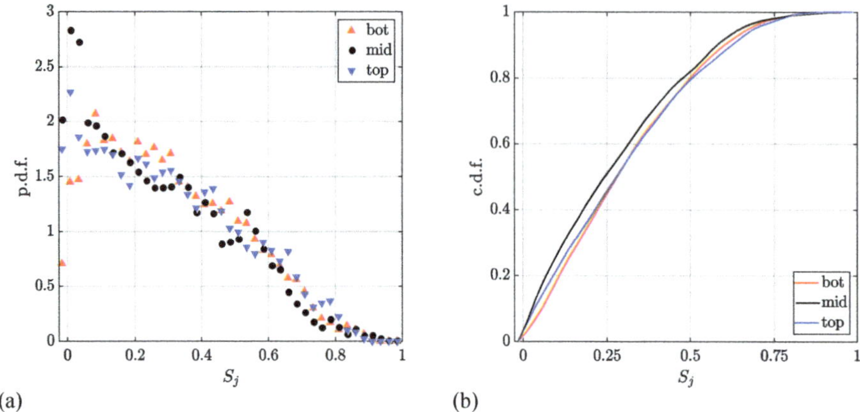

Figure 5.26: Probability distributions of the LSC strength at the bottom ($x/L = 0.25$), middle ($x/L = 0.5$) and top ($x/L = 0.75$) of the cell: (a) probability density function; (b) cumulative density function. Data from the numerical simulations.

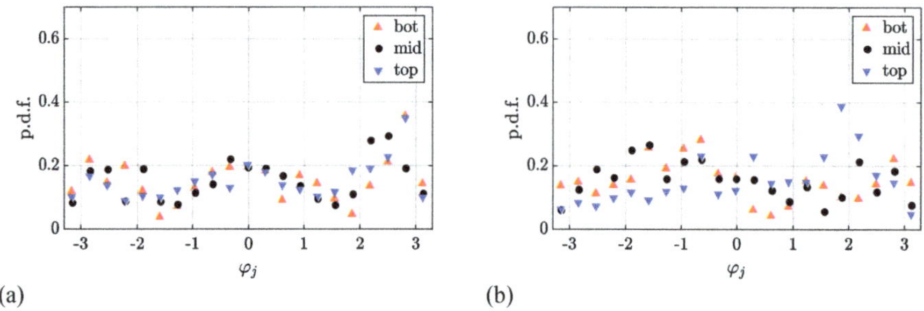

Figure 5.27: Probability distributions for the LSC orientation at the bottom ($x/L = 0.25$), middle ($x/L = 0.5$) and top ($x/L = 0.75$) of the cell for (a) the SRS and (b) the DRS. The angles are in radians. Data from the numerical simulations.

top plate and the relatively high thermal resistance of the latter.

A second discrepancy concerns the statistical distribution of the LSC orientations, reported for the numerical case in figure 5.27. Both the distributions for the SRS (figure 5.27a) and the DRS (figure 5.27b) do not present any peak, representative of a preferential angular orientation. This is a predictable result since, on one side, the numerical setup is devoid of imperfections, on the other, some secondary effects, like the Earth's Coriolis force, are not considered in the simulation.

Finally, a further considerable difference between experimental and numerical findings regards the distribution of the angular velocity of the LSC. The p.d.f.s of the numerical case (figure 5.28) are much narrower than the experimental counterparts, both when all the samples are used and when the statistics are restricted to the samples with $S_m > 0.5$ or $A_m > \overline{A_m}$. Indeed, while in the experimental case the p.d.f. related to the entire set of samples has a quasi Gaussian shape, in the numerical

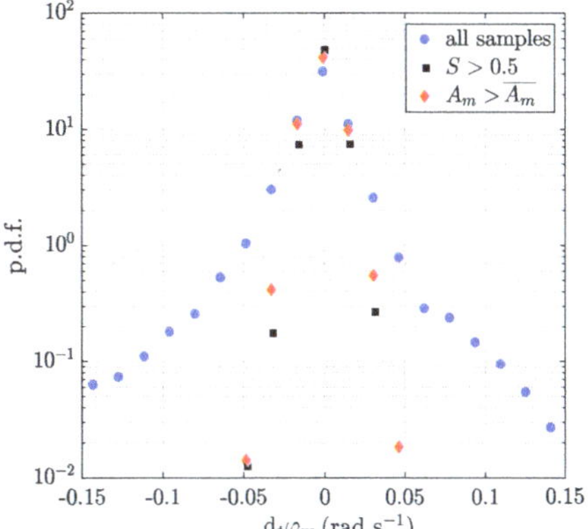

Figure 5.28: Probability distribution for the angular velocity of the LSC. Data from the numerical simulations.

case it exhibits an exponential distribution. However, the p.d.f.s for the subsets of samples $S_m > 0.5$ or $A_m > \overline{A_m}$ preserve a Gaussian shape, but with a lower standard deviation.

## 5.5. Conclusions

THE present chapter has focused on the analysis of RB convection in non-rotating conditions at $Ra = 1.86 \times 10^8$ and $Pr = 7.6$ using both experiments and DNS. The characteristic modes of the turbulent convection have been identified by means of modal decomposition techniques, namely POD for experimental data and extended POD for numerical results. The statistical behavior of the LSC has been characterized with conventional diagnostics methods based on the analysis of the azimuthal vertical velocity profiles near the cylinder sidewall at different heights. Results consistent with the existing literature on the topic have been retrieved experimentally, while an improved method to identify the different flow states (SRS and DRS) of the LSC exploiting the low-order POD modes has been introduced. Simulations including the effects of the properties of the cylinder sidewall show excellent agreement with the experimental measurements and confirm the relevant influence of the temperature condition of the sidewall itself on the flow dynamics.

The main findings of the present investigation can be summarized as follows:

- the mean flow field exhibits azimuthal symmetry, irrespective of the actual temperature condition at the external side of the cylinder sidewall (as long as it is uniform); the latter can anyhow introduce an asymmetry between the bottom and the top halves of the cylindrical sample, affecting the dynamics of rising hot and sinking cold plumes;

- the first POD modes of the fluctuating velocity field determined experimentally capture the principal states of the LSC and appear to be coupled in pairs, as a direct consequence of the statistical axisymmetry of the flow;

- the first two POD modes of the temperature fluctuation determined numerically does not feature clearly a state of the LSC and could be associated with the transitional states of the same;

- identification of the LSC orientation based on POD modes seems to be devoid of any bias, differently from methods based on the analysis of azimuthal profiles of temperature or vertical velocity;

- statistics show the absence of a preferential state of the LSC, confirming the chaotic behavior of the investigated flow.

# 6

# Investigation of rotating Rayleigh-Bénard convection

$I$N this chapter a comparison between the results of the experimental and numerical investigation of RB convection with background rotation is carried out. The experimental measurements are performed for two different Rossby numbers, namely $Ro = 0.25$ and $Ro = 0.1$, at a Rayleigh number equal to $2.86 \times 10^8$ and Prandtl number equal to $6.4$. These parameters correspond to an average temperature of the water of $23.5\,°C$ and a temperature difference still equal to $5\,°C$, while the rotation rate is $0.523$ rad/s and $1.308$ rad/s for the highest and the lowest Rossby numbers, respectively. Differently from the non-rotating experiments, the temperature of the tank water is controlled and maintained constant at $23.5\,°C$ by means of the Peltier cooling system driven by the TEC. This prevents the occurrence of asymmetries of the flow field related to the specific thermal condition of the cylinder sidewall (see discussion in section 5.1).

In each case, measurements are performed over four hours after having reached a steady state. Rotation is introduced only after the temperature gradient has been set and the system has been adapted to this condition. After addition of rotation, the system is allowed to adapt to the new dynamical condition for at least one hour. This time is very conservative since, following the reasoning of Kunnen $et$ $al.$ [125], the settling time for impulsive spin-up is the Ekman time scale $\tau_E = L/\sqrt{\nu\Omega}$, which at the lowest rotation rate, among the two investigated ones, is equal to $212\,$s. Moreover, such a time is further reduced by the presence of the convective motion before the rotation is applied.

Below, some results from the experimental investigation are presented and compared to the output of the numerical simulations. The comparison focuses essentially on the time-averaged velocity fields and the main aspects of the statistical behavior of the coherent structures.

The following discussion is organized in a different way from the previous chapter. Specifically, we first introduce the results of the numerical simulations of rotating RB convection to provide the reader with a picture of what is expected with varying the background rotation. The numerical simulations are performed in the same conditions of $Ra$ and $Pr$ of the non-rotating experiment and for several $Ro$, among which are those investigated experimentally (see Table 4.1 for complete information about the simulation parameter settings). The procedure of the numerical simulations is basically the same as the experiments: first, RB convection is simulated in absence of the background rotation for about 350 free-fall time scales; subsequently, rotation is introduced and the simulation is runned at least for 1,000 free-fall time scales. In some cases, the simulation time is extended up to more than 3,000 time units to be sure about the convergence of the results. It is here remarked that the simulations include the presence of the physical sidewall and the thermal condition imposed at its external side is isothermal at the average fluid temperature. In the second part of this chapter, we comparatively present the experimental measurements and discuss the source of analogies and differences with the numerical simulations.

## 6.1. Effects of Rossby number on the mean flow characteristics

### 6.1.1. Influence on the velocity field

Figure 6.1 reports the time and azimuthally averaged fields of the velocity components $u_x$ (axial), $u_r$ (radial) and $u_\varphi$ (azimuthal) and of the temperature $\theta$ for four different Rossby numbers, namely $Ro = \infty$, 1, 0.25 and 0.1.

The non-rotating configuration (figure 6.1a-d) is here reported again as a reference. The structure of the mean flow field in this case has been already described in the previous chapter. The tendency to axisymmetry results in very small azimuthal velocities in the time-average evolution as confirmed by the map in figure 6.1c. From figure 6.1 it can be noted that the radial velocities are greater in the regions near the top and the bottom plates than in the middle part of the cylinder. This behavior is essentially related to the dynamics of the plumes which, in the vicinity of the plates, are driven radially outward by the LSC and convected along the sidewall. This is also clear from the temperature map in 6.1d, where it can be seen that the thermal boundary layers at the plates are stretched along the lateral wall, while the temperature of the bulk is uniform and equal to the fluid average temperature.

The introduction of a background rotation causes the development of mean azimuthal velocities, while the axial and radial velocities are observed to decrease with respect to the non-rotating case. For $Ro = 1$ (figure 6.1e-h) the recirculations adjacent to the plates are squeezed towards the sidewall and the mean axial velocities in correspondence of the cylinder axis are considerably reduced. Anticyclonic motion (i.e., azimuthal rotation in the direction opposite to the background rotation) is observed in the regions adjacent to the plates, while cyclonic motion develops in the central part of the convection cell near the sidewall. As motivated by Kunnen *et al.* [71], the azimuthal motion is due to the action of the Coriolis forces that tend to deflect the flow with horizontal velocity to the right with respect to the local direction of the horizontal velocity itself. Therefore, the radially outward flow in proximity of the plates results in an anticyclonic motion, whereas the weak radial inward in the

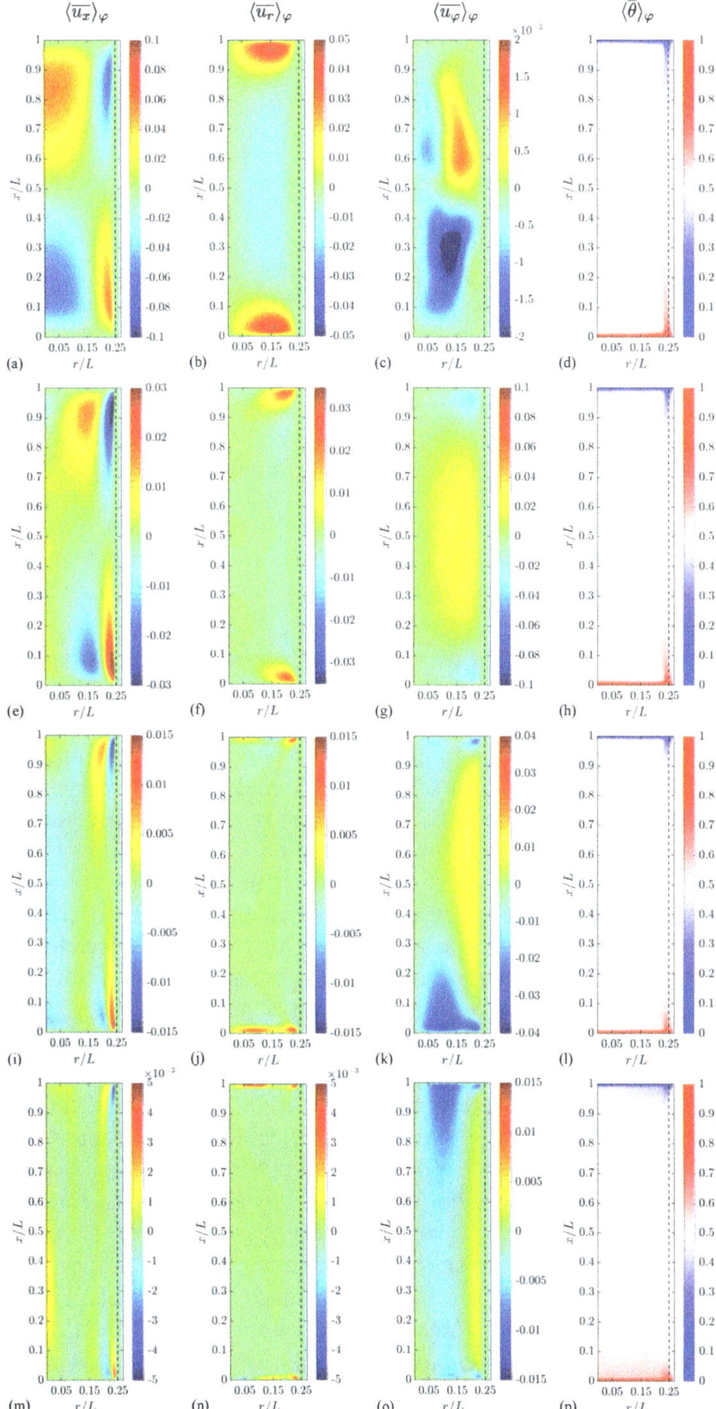

Figure 6.1: See next page for the caption.

central part of the cell induces cyclonic rotations in these regions. Such a mechanism could be also explained with conservation of the angular momentum, according to which a radial outward motion has to correspond to a decrease of the azimuthal velocity, that is to say to anticyclonic motion, and viceversa. It is worth noting that $Ro = 1$ is in the second regime of the rotation according to the classification reported in section 1.5. As a consequence, the instantaneous evolution is characterized by the occurrence of columnar vortices and a breakdown of the LSC. Indeed, the presence of the two recirculations adjacent to the sidewall in the mean flow field is clearly a legacy of the LSC, while the decrease of the mean vertical velocity along the cylinder axis can be associated with the emergence of the columnar structures in this region of the velocity field.

Moving to $Ro = 0.25$ (higher rotation rates), on one side the flow velocities are observed to further decrease, on the other side the structure of the flow field is found to substantially change. Specifically, in both the lower and the upper part of the convection cell it is possible to detect two recirculations, a primary one in the neighborhood of the sidewall ($0.15 < r/L < 0.25$) and a secondary one in the central part of the convection cell ($r/L < 0.15$). The latter is essentially driven by the Ekman suction mechanism; note, in fact, that the mean flow is directed towards the plates in the proximity of the cylinder axis (figure 6.1i). The radial outward velocity in the Ekman layer (figure 6.1j) is associated with anticyclonic motion (figure 6.1k), while in the middle of the cylinder sample no cyclonic motion is observed. As concerns the primary circulation close to the sidewall, it is reminiscent of the recirculation due to the LSC of the RB convection at slower or no rotation rate. Indeed, such a circulation is still related to the convection of plumes which are not affected by the Ekman suction in this region of the flow field. As observed at higher Rossby number, this circulation is characterized by anticyclonic motion in the proximity of the plates and cyclonic motion in the middle part, which can be always explained by consideration of the Coriolis forces or the conservation of the angular momentum. It should be mentioned that the boundary layer developing on the lateral sidewall is typically referred to as Stewartson boundary layer [203–207]. As first shown by Stewartson [203], when the angular velocity of the bulk fluid is discontinuous along the radius, this layer has a sandwich structure consisting of an outer layer of thickness $O(\nu^{1/4})$ and a thinner inner layer of thickness $O(\nu^{1/3})$. In non-dimensional terms, the thickness of these two layers are equal to $Ek^{1/4}$ and $Ek^{1/3}$, respectively, where $Ek = \nu/(\Omega L^2)$ is the Ekman number. Note that, differently from the Rossby number, the Ekman number does not depend on the buoyancy properties of the system. In the present case, $Ek = 8.07 \times 10^{-5}$ and the non-dimensional thicknesses of these layers are $0.095$ and $0.043$, respectively. Such values are consistent with the extent of the primary recirculation zone adjacent the sidewall: more specifically, the inner Stewartson layer corresponds to the vertical drift of the flow away from the plates in the very proximity of the sidewall, whereas the outer layer corresponds to the region between $r/L \approx 0.17$ and $r/L \approx 0.22$ where the flow is returned.

A remarkable observation about the flow field at $Ro = 0.25$ is that the distribution of $\langle \overline{u_\varphi} \rangle_\varphi$ (figure 6.1k) is not symmetric between the lower and the upper part of the convection cell. Indeed, a stronger anticyclonic motion is found in the proximity of the bottom plate. Being the governing equations solved by the numerical code invariant under a reflection about the mid-plane of the convection cell, a symmetry of the flow field is indeed expected; hence, the results shown in figure 6.1k suggest that a greater number of iterations is needed for convergence in this operating conditions. The specific considered simulation has been runned for 3,500 non-dimensional times after the onset of rotation; minor differences have been found between the cumulative averages over the last 1,000 time units and also the Nusselt number is found to be converged within 1%. Therefore, we can only expect that the simulation is reproducing an unsteady phase of the flow evolution characterized, on the average, by the top/bottom asymmetry of figure 6.1k that persists over a considerably long time. After all, the flow velocities are significantly lower (at least one order of magnitude) than those corresponding to the non-rotating cases and longer characteristics times of the dynamical evolution are indeed expected. Note that the pointed out asymmetry is visible also in the temporal and azimuthal averages of the other thermo-fluid dynamic quantities.

When further increasing the rotation rate ($Ro = 0.1$, figures 6.1m-p), velocities are further reduced and the mean flow tends to organize in two distinct regions: at the center of the cell, around the cylinder axis, anticyclonic motion is established, while near the sidewall the circulation associated with the Ekman and Stewartson layers is still observable, although it is confined in a narrower region due to the increased Ekman number. It is interesting to note that the mean secondary circulation observed in the bulk for $Ro = 0.25$ driven by the Ekman suction is not distinguishable for $Ro = 0.1$ and replaced by an irregular pattern (see central part of the map in figures 6.1m), which results from temporal and azimuthal averaging of the Taylor columns inhabiting the cylinder interior at this low Rossby number. Finally, it is possible to observe that the temperature field is much more diffused than at higher Rossby numbers.

## 6.1.2. Influence on the heat transfer

THE analysis carried out in the previous section suggests that by decreasing the Rossby number the mean flow switches from a circulatory convective stream to a nearly anticyclonic co-rotating flow, although a recirculation associated with the Ekman and the Stewartson layer remains near the sidewall. As explained in the introduction of the present thesis, such changes in the flow field affect significantly also the heat transfer. To show the above effect, figure 6.2 reports the behavior of the Nusselt number as a function of the Rossby number computed from the numerical

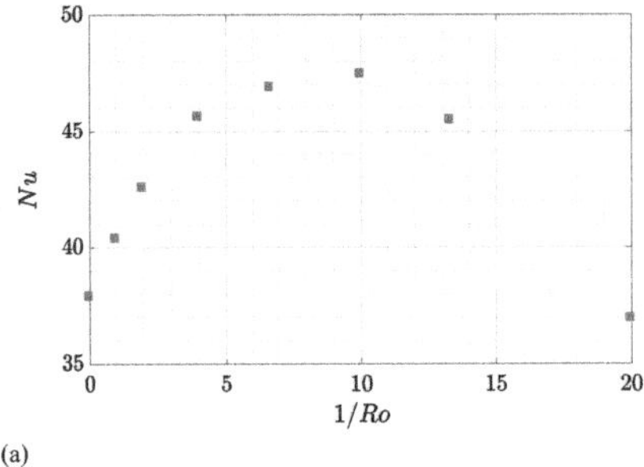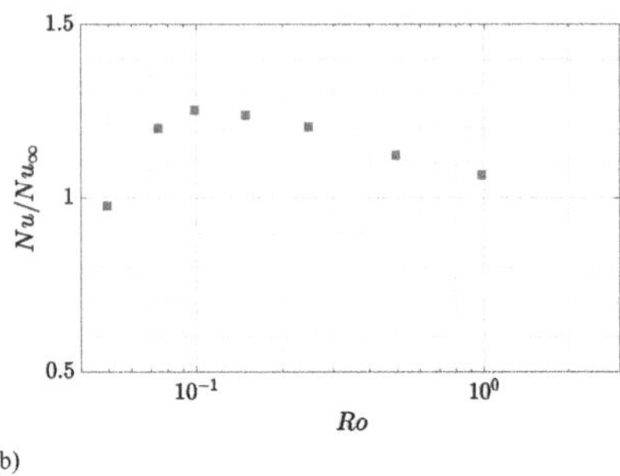





Figure 6.2: Heat transfer in rotating RB convection at $Ra = 1.86 \times 10^8$ and $Pr = 7.6$ in a cylinder with aspect ratio $\Gamma = 1/2$: (a) behavior of $Nu$ as a function of the inverse Rossby number $1/Ro$; (b) enhancement of the heat transfer in presence of the rotation with respect to the non-rotating case. Simulations include the effects of the physical sidewall, which is identical to that of the present experimental setup $((\rho C_p)_s/(\rho C_p)_f = 0.5$, $\kappa_s/\kappa_f = 0.32$ and $t_s/L = 0.0207)$, with uniform temperature $\theta_e = 0.5$ at its external side.

simulations; here, for reasons of clearness, we report the results corresponding not only to the cases presented and discussed in figure 6.1 but also to other simulations performed at the same $Ra$ and $Pr$, always with the presence of the physical sidewall. It is possible to see that for the considered $Ra$ and $Pr$ the maximum enhancement of the Nusselt number with respect to the non-rotating case is achieved at $Ro = 0.1$, while both $Ro = 0.25$ and $Ro = 1$ are in the second regime of the rotating thermal convection.

It is worth noting that the Nusselt number in figure 6.2 has been calculated as mean between the non-dimensional area-averaged heat flux entering the cell through the bottom plate and that leaving it from the top plate. Such heat fluxes are in general different from the heat flux through the generic horizontal section of the sample since locally there may be a heat flux exchanged between the fluid and the sidewall. Figure 6.3 shows the axial profile of the time- and azimuthally-averaged Nusselt number through the interface fluid/sidewall for the investigated Rossby numbers. Such a Nusselt number is defined as the lateral heat flux $\dot{q}_s$ normalized by the pure conductive heat flux $\kappa\Delta/L$ and, in the non-dimensional terms, it is simply calculated as the radial derivative of the fluid temperature at the interface. For continuity of the heat flux, this is also equal to the radial derivative of the sidewall temperature at the interface multiplied by the ratio of the thermal conductivities of the fluid and the wall. From figure 6.3 it is evident that by decreasing the Rossby number the amount of the heat flux locally exchanged between the fluid and the sidewall essentially decreases up to $Ro = 0.1$, then it starts increasing slightly. Such an increase corresponds to the decrease of the Nusselt number in this regime, which is that fully dominated by the rotation. However, it can be noted the total heat flux across the interface fluid/sidewall is zero, since the the heat current leaving the fluid in the bottom half of the cell is equal to the heat current entering the fluid in the top half. This means that this exchange does not affect the global heat transfer from the bottom to the top, i.e. the Nusselt number.

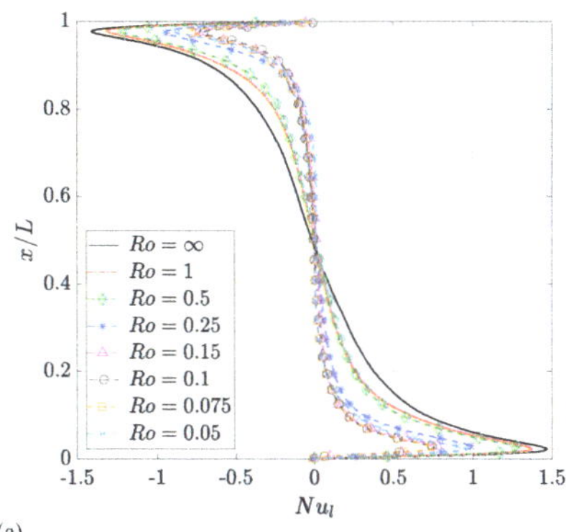 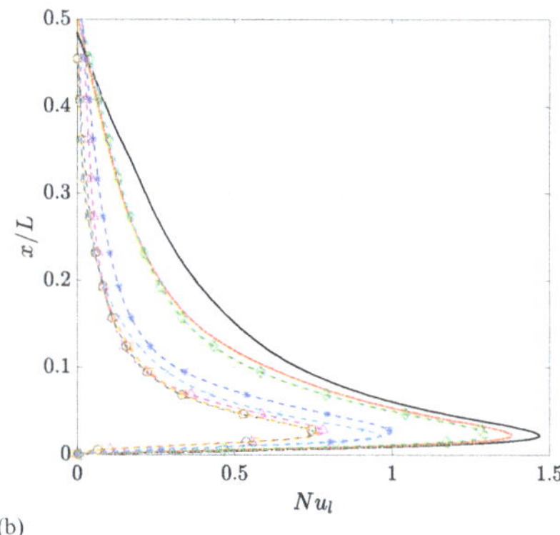

Figure 6.3: Heat flux through the lateral wall in the sample with the physical sidewall with $\theta_e = 0.5$ and $Ra = 1.86 \times 10^8$, $Pr = 7.6$ and $\Gamma = 1/2$: (a) axial profile of the time- and azimuthally-averaged Nusselt number through the interface fluid/sidewall; (b) zoom in the region $0 \leq x/L \leq 0.5$. Results from numerical simulations.

To provide further insight into the mechanisms determining the behavior of the Nusselt number observed in figure 6.2, the radial profile of the local Nusselt number through the bottom plate is reported in figure 6.4. The diagram shows that the increase of the Nusselt number at moderate rotation rate ($Ro = 1$ and $Ro = 0.5$) is essentially related to an increase of the heat transfer through the bottom plate in proximity of the sidewall; indeed, the heat transfer in the central region is found to be smaller than that corresponding to the non-rotating case ($Ro = \infty$). The latter behavior can be explained considering that in these conditions the Ekman suction/pumping mechanism is not so strong to induce the formation of the columnar vortices that would lead to an increase of the heat transfer in the central part of the cell, but it surely affects the dynamics of the plumes. The increased Nusselt number in the region $r/L > 0.13$ is indeed related to a better heat transfer efficiency of the plumes that detach at this location. However, the enhancement of $Nu$ with decreasing $Ro$ up to $Ro = 0.1$ is also related to the increase of the local Nusselt number in the central part of the cell, which is due to the emergence of the Taylor columnar vortices.

Entering the third regime ($Ro > 0.1$ in the present case), we observe an opposite evolution that causes the decrease of the global Nusselt number. At $Ro = 0.075$, the local Nusselt number through the bottom plate is nearly unchanged in the central part of the cell, whereas a substantial drop is found for $r/L > 0.15$. The explanation for this might be a weakening of the circulation associated with the Ekman and the Stewartson boundary layers. When the Rossby is further reduced, the local Nusselt number diminishes both in the center and the near-sidewall part of the cylindrical sample.

In conclusion, the above analysis shows that the recirculations related to the Ekman and the Stewartson boundary layers play an important role in the variation of the heat transfer with rotation. From figure 6.4 it is evident that the major difference

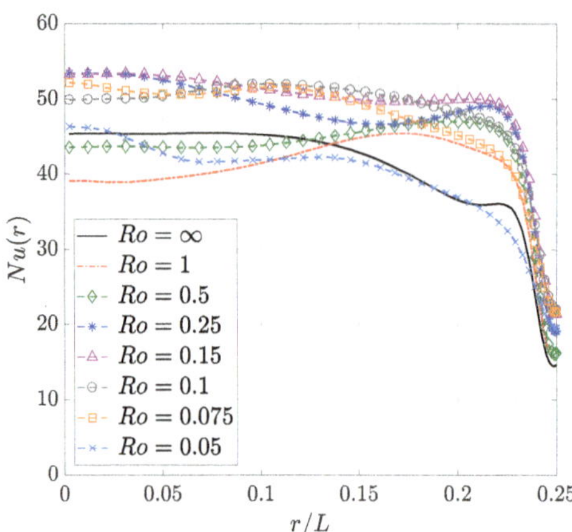


Figure 6.4: Heat flux through the bottom plate in the sample with the physical sidewall with $\theta_e = 0.5$ and $Ra = 1.86 \times 10^8$, $Pr = 7.6$ and $\Gamma = 1/2$. Results from the numerical simulations.

in the local heat transfer from the bottom plate are observed close to the sidewall. Interestingly, the increase of local vertical $Nu$ with decreasing $Ro$ corresponds to a decrease of the lateral $Nu$ through the sidewall (figure 6.3). This suggests that the flow is less sensitive to the presence of the sidewall in such conditions. However, further investigation on this point is needed to elucidate the actual mechanisms leading to the above-discussed behavior in the time-averaged evolution.

## 6.2. Experimental measurements and comparison with simulations

I N this section, experimental measurements of the rotating RBC are presented and compared with the numerical simulations reported in the previous section. The slightly different values of $Ra$ and $Pr$ between the experiments and the simulations do not constitute a source of significant differences, at least from the point of view of a qualitative comparison, which is the main focus of the present discussion.

Figure 6.5 reports the time-averaged maps of the axial, radial and azimuthal velocity components for the two different investigated $Ro$. Such maps have been obtained by averaging about 480 snapshots, one for each half minute of the experiment, therefore they represent ensemble averages rather than a temporal ones. In addition, the velocity fields have been azimuthally averaged to allow an analysis of the different behaviors in the central part and the near-sidewall region of the flow field, in analogy with the discussion about the numerical results. It is immediately evident that the measured velocity fields have very different structures from those found in the numerical simulations at both the investigated Rossby numbers.

Focusing on $Ro = 0.25$ (figure 6.5a-c), it is possible to see that the radial velocity

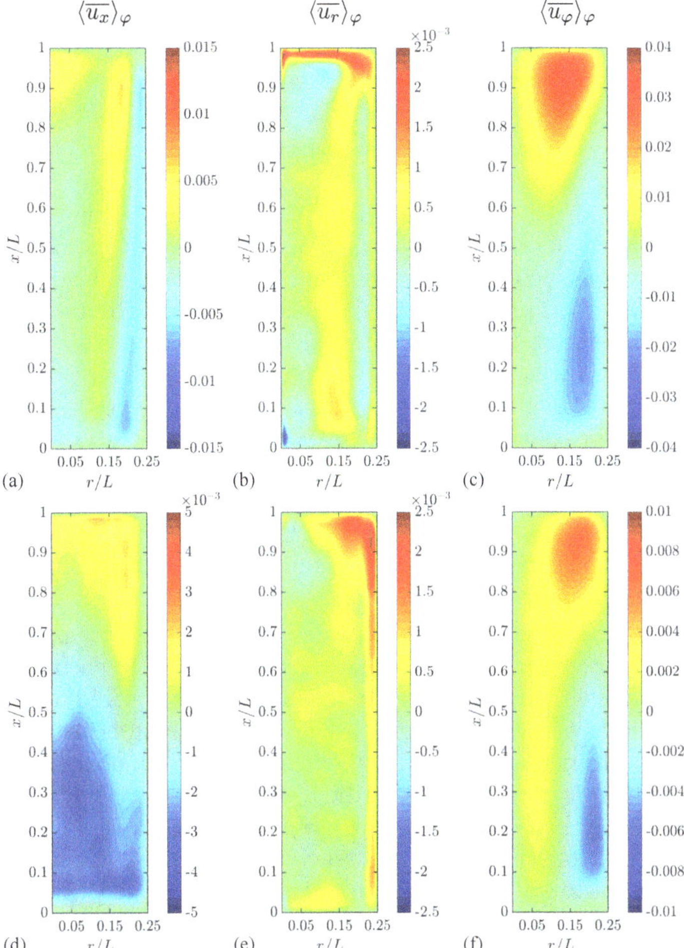


Figure 6.5: Experimental measurements of the rotating convection for $Ra = 2.86 \times 10^8$ and $Pr = 6.4$. From left to right, time- and azimuthally-averaged fields of the vertical, radial and azimuthal velocity components. The rows from top to bottom correspond to (a)-(c) $Ro = 0.25$ and (d)-(f) $Ro = 0.1$. Velocities are scaled by the free-fall velocity.

component is significantly smaller than the axial and the tangential ones. This explains the absence of a large-scale coherent pattern, although considerably high radial outward velocities are found near the top plate. From the inspection of the vertical velocity field it is possible to detect a recirculation in the region adjacent to the sidewall with downward velocities close to sidewall and positive velocities at smaller radial distances from the cylinder axis. Such a recirculation is different from that related to the Ekman and Stewartson boundary layers observed in figure 6.1i always in the region close to the sidewall. The latter indeed consists of two separate recirculations extending over one half of the convective cell, whereas that observed in the experiments covers the entire height. It should be noted that however such a recirculation is not planar since it is associated with the azimuthal velocities shown in figure 6.5c. In the central part of the convection cell vertical flow towards the plates is observed, probably due to the Ekman suction mechanism.

The major discrepancies between the numerical simulations and the experiments are undoubtedly found in the distribution of the azimuthal velocities (compare figure 6.1k and figure 6.5c). As discussed before, in the numerical simulation cyclonic rotation is observed close to the sidewall, while anticyclonic rotation takes place in the central part near the plates. This radial separation between cyclonic and anticyclonic flows is not present in the flow field measured experimentally; on the contrary, it is evident that the lower part of the fluid layer exhibits anticyclonic rotation, while the flow in the upper half of the cell is cyclonic. It is here noted that in the experiment the angular velocity vector is parallel to the gravity, therefore the anticyclonic velocities add to the background rotation, while the cyclonic ones are subtracted from it. If the Coriolis forces are invoked to explain the genesis of such azimuthal velocities, radial outward velocities should correspond to the cyclonic (positive) azimuthal flow, while radial inward flow would result in anticyclonic (negative) azimuthal velocities. This seems quite consistent with the flow near the plates, but it seems not to hold for the flow in the bulk.

At the lower Rossby number ($Ro = 0.1$), smaller velocities are found in agreement with the observations drawn from numerical simulations. In particular, both the axial and the radial velocities are found to be very small compared to the azimuthal velocities (about one order of magnitude). The distribution of the azimuthal velocity is not so different from that observed at the higher Rossby number ($Ro = 0.25$): we can detect again anticyclonic motion in the lower half of the cylinder and cyclonic motion in the upper half. However, at a closer look it is possible to see that the central part of the convection cell is characterized mainly by positive azimuthal velocities (opposite to the background rotation). Such a behavior resembles that observed in the numerical case, where an anticyclonic rotation (again opposite to the background rotation) is established in the mean flow field in the central part of the cell.

In Figures 6.6 and 6.7, the azimuthal averages of the root mean square of the vertical, radial and azimuthal velocity fluctuations corresponding to $Ro = 0.25$ and $Ro = 0.1$ are reported for the experiment and the numerical simulations, respectively. These figures confirm the result already observed for the non-rotating case that the experimentally measured fluctuations are on the average smaller than those calculated numerically. In both experiments and numerical simulations and in both the cases $Ro = 0.25$ and $Ro = 0.1$, significant vertical velocity fluctuations are found in the middle part of the convection cell far from the base plates. Moreover, the largest fluctuations of this velocity components are reached in the proximity of the wall, while slightly lower values are observed around the cylinder axis. Conversely, as concerns the average fluctuations of the radial and tangential velocity components, significant fluctuations are observed in the region close to the base plates, while nearly zero fluctuations are found in the bulk of the fluid.

In this regard, it should be noted that the fluctuations of the velocity components

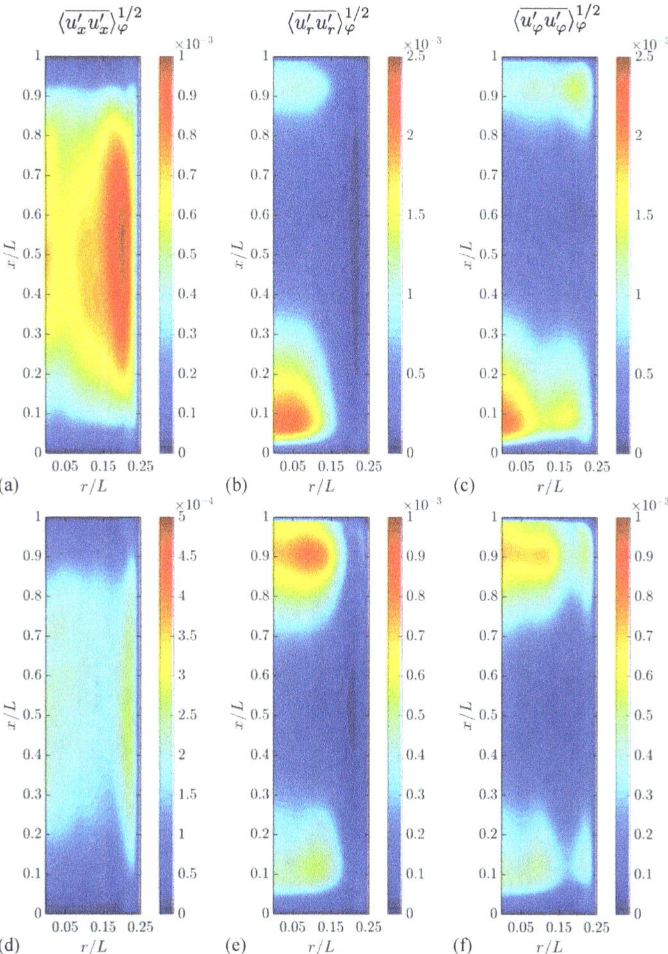



Figure 6.6: Experimental measurements of the rotating convection for $Ra = 2.86 \times 10^8$ and $Pr = 6.4$. From left to right, azimuthally-averaged fields of the root mean square of the vertical, radial and azimuthal velocity fluctuations. The rows from top to bottom correspond to (a)-(c) $Ro = 0.25$ and (d)-(f) $Ro = 0.1$. Velocities are scaled by the free-fall velocity.

in the proximity of both the plates and the sidewall are affected by fairly high uncertainty in the experimental case, definitely greater than that related to the non-rotating experiments present in the previous chapter. This is essentially related to the fact that the seeding density in the rotating experiments is lower because of the faster sedimentation of the particles due to the presence of rotation, as explained in section 2.2 and proved by several investigations ([140] to cite one). The lower seeding density implies that in the transformation from Lagrangian to Eulerian reference frame, with the approach used in the current investigation (see section 3.1.4), based on interpolation of sparse data on a structured grid, a larger interrogation volume has

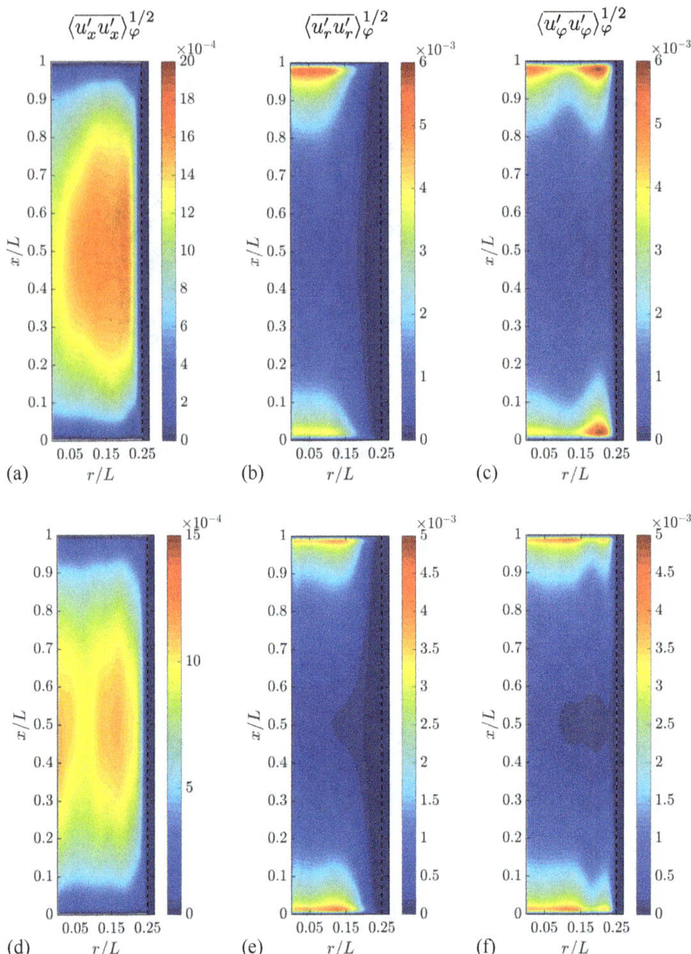

Figure 6.7: Numerical results of the rotating convection for $Ra = 1.86 \times 10^8$ and $Pr = 7.6$. From left to right, azimuthally-averaged fields of the root mean square of the vertical, radial and azimuthal velocity fluctuations. The rows from top to bottom correspond to (a)-(c) $Ro = 0.25$ and (d)-(f) $Ro = 0.1$. Velocities are scaled by the free-fall velocity.

to be used to estimate the velocity in a selected grid point. In fact, the interrogation volume should be so large that a sufficient number of tracked particles falls in it; the minimum number required depends on the kind of polynomial fitting function used for interpolation and is 10 for a second-order polynomial as that employed here. However, the effective number of particles to be used should be at least equal to 1.5-2 times the minimum number required to avoid that outlier velocity data (due to reconstruction errors or residual ghost particles with motion different from the local flow) could affect significantly the velocity estimation itself. Such requirements are even more severe for the grid points close to the sidewall and the plates, since, on

 | CHAPTER 6. Investigation of rotating Rayleigh-Bénard convection

one side, the interrogation volume does not fall entirely in the measurement domain and, correspondingly, a greater size of the interrogation volume is required, on the other and most critical side, the presence of kinematic boundary layers at the wall results in large velocity gradients, which could deteriorate the goodness of the fit as the interrogation volume size is increased. Therefore, a compromise is definitely mandatory.

Following the above reasoning, it should be considered that both the simulations presented in section 6.1 and theory [203–207] demonstrates that with increasing the rotation rate, the thicknesses of the Ekman and the Stewartson layers forming on the plates and the sidewall decrease roughly as $Ek^{1/2}$ for the Ekman layer and $Ek^{1/4}$ and $Ek^{1/3}$ for the inner and the outer Stewartson layers, respectively. This means that with increasing the rotation rate a finer spatial resolution is indeed required. However, the faster particle sedimentation goes against such a requirement. Indeed, at a closer look, in the case $Ro = 0.1$ the relatively high vertical velocity fluctuations (figure 6.6d) near the sidewall could also be related to the error in the particle velocity interpolation in this region.

By comparing the fields of the radial and azimuthal velocity fluctuations related to the cases $Ro = 0.25$ and $Ro = 0.1$, it is possible to see that the highest values of both the radial and azimuthal velocity fluctuations are achieved in correspondence of the lower plate in the former case and of the upper plate in the latter. Indeed, we do not have a plausible explanation of this feature (symmetry between the bottom and the top halves is expected in both cases) and further investigation is required to elucidate the origin of such a discrepancy.

Ultimately, we can conclude that poor agreement is found between experiments and numerical simulations as concerns the mean velocity fields. Indeed, relevant differences are found also in the instantaneous velocity fields. As an example, figure 6.8 reports a comparison between the instantaneous velocity fields of the numerical simulations and the experiments for both the case $Ro = 0.25$ and the case $Ro = 0.1$. The large-scale coherent structures are here identified by the $Q$-criterion; the same $Q$-threshold is used for both the numerical and experimental cases. As regards $Ro = 0.25$, the coherent structures of the flow field measured experimentally (figure 6.8c) have the appearance of thick columnar structures with a slightly twisted shape. In the numerical flow field at the same $Ro$ (although slightly different $Ra$ and $Pr$), the coherent vortical structures are surely thinner and their number appears to be higher. At $Ro = 0.1$, the shape of the coherent structures looks similar between experiments and simulations, but again the density of these structures is evidently greater in the numerical case. Note that, in the experimental case, the derivative in the region close to the plates are not estimated with sufficient accuracy due to the limitations of spatial resolution. This is the reason why no structures are visible in such a region in figure 6.8c and figure 6.8d with the prescribed threshold of $Q$.

It is important to remark that the comparison of the dynamical behavior of the flow observed in the experiment and in the simulation based on the analysis of the instantaneous flow field is not so straightforward. Therefore, following the investigation line of the previous chapter, the POD modes of the fluctuating velocity field are presented below. Figure 6.9 shows the first two POD modes for the case

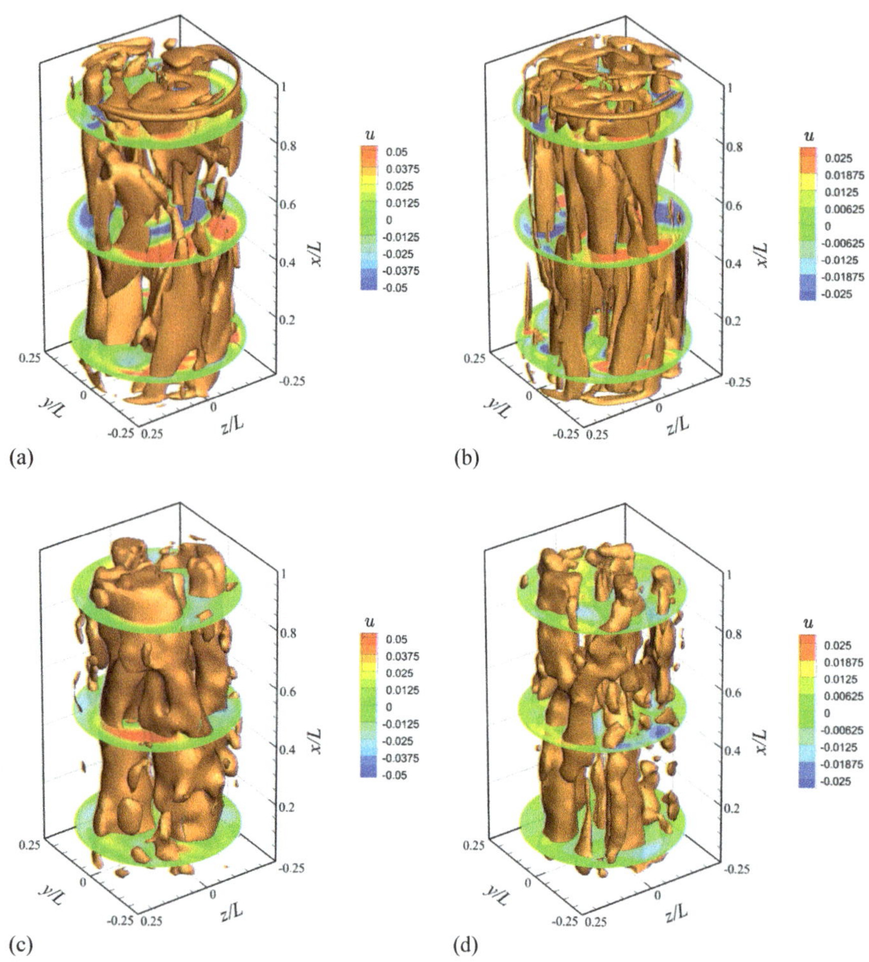

Figure 6.8: Comparison between numerical simulations and experimental results. Instantaneous velocity fields for: (a) $Ra = 1.86 \times 10^8$, $Pr = 7.6$ and $Ro = 0.25$, numerical simulation; (b) $Ra = 1.86 \times 10^8$, $Pr = 7.6$ and $Ro = 0.1$, numerical simulation; (c) $Ra = 2.86 \times 10^8$, $Pr = 6.4$ and $Ro = 0.25$, experiment; (d) $Ra = 2.86 \times 10^8$, $Pr = 6.4$ and $Ro = 0.1$, experiment. Slices of the vertical velocity component field at three different heights ($0.1L$, $0.5L$ and $0.9L$) and isosurfaces of $Q$. Velocities are scaled by the free-fall velocity.

$Ro = 0.25$ for both the numerical simulation and the experiment, represented through a set of horizontal slices of the vertical velocity field. Indeed, these modes appear to have a strikingly similar flow topology between the experiments and the numerical simulations especially in the middle part of the convection cell. The pattern observed consists of two vertical opposite currents with a twisted shape that resembles a LSC circulation; however, the analysis of the azimuthal velocities of these modes (not reported here for brevity) reveals that these currents are characterized by azimuthal velocities which are opposite in the bottom and the top half of the convection cell.

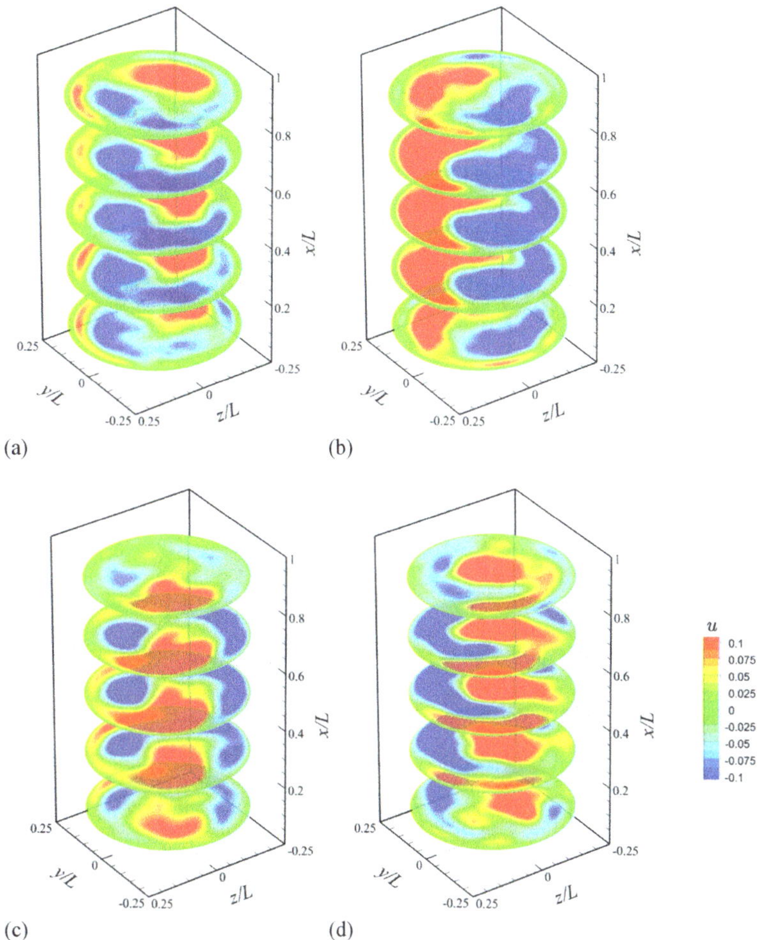

Figure 6.9: Comparison between the POD modes of the fluctuating velocity field for $Ro = 0.25$;. (a)-(b) 1st and 2nd POD modes from the numerical simulation at $Ra = 1.86 \times 10^8$ and $Pr = 7.6$; (c)-(d) 1st and 2nd POD modes from the experiment at $Ra = 2.86 \times 10^8$ and $Pr = 6.4$. Slices of the vertical velocity component at five different heights ($0.1L$, $0.3L$, $0.5L$, $0.7L$ and $0.9L$). Velocities are scaled by the maximum value of the velocity magnitude for each mode.

In particular, the 1st numerical mode (figure 6.9a) and the 2nd experimental mode (figure 6.9d) are characterized by cyclonic (anticyclonic) rotation in the upper part of the domain and anticyclonic (cyclonic) motion in the lower half. Conversely, in the other modes the in-plane motion is characterized by two adjacent opposite circulations that switches their direction passing through the top to the bottom part of the cell; thus, a quite complex flow geometry is indeed observed in such cases.

The 3rd and the 4th mode calculated from the numerical runs and the experiments are instead reported in figure 6.10. Structural similarities between the 3rd numerical mode (figure 6.10a) and the 4th experimental one (figure 6.10d), on one side, and

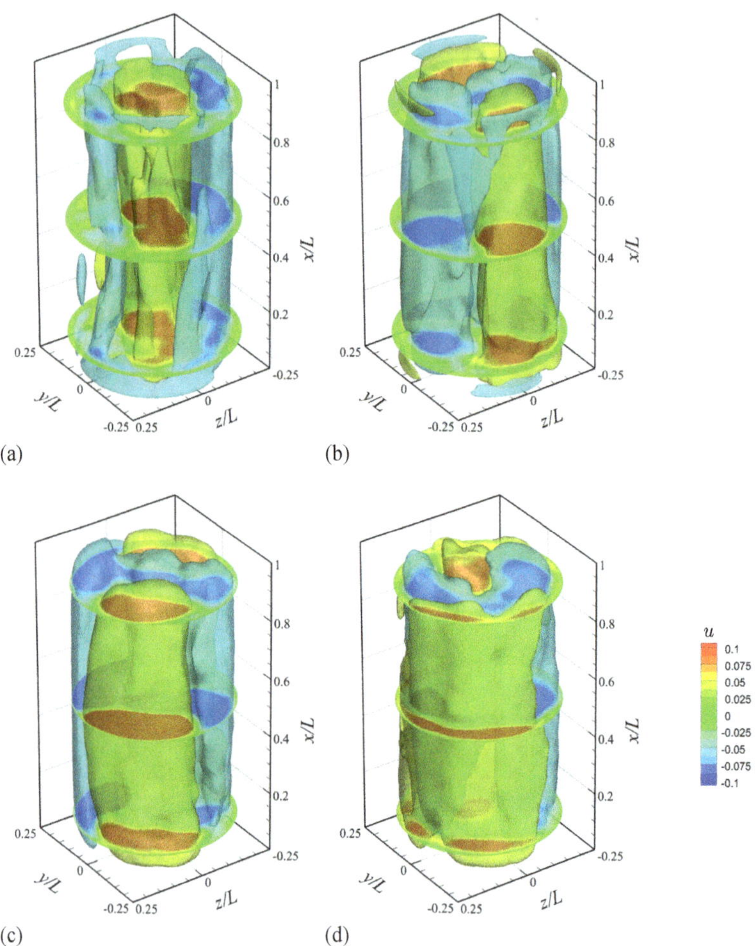

Figure 6.10: Comparison between the POD modes of the fluctuating velocity field for $Ro = 0.25$;. (a)-(b) 3rd and 4th POD modes from the numerical simulation at $Ra = 1.86 \times 10^8$ and $Pr = 7.6$; (c)-(d) 3rd and 4th POD modes from the experiment at $Ra = 2.86 \times 10^8$ and $Pr = 6.4$. Slices of the vertical velocity component at three different heights ($0.1L$, $0.5L$ and $0.9L$) and isosurfaces of positive and negative vertical velocity. Velocities are scaled by the maximum value of the velocity magnitude for each mode.

between the 4th numerical mode (figure 6.10b) and the 3rd experimental one (figure 6.10c) can be indeed noticed. In particular, the latter two modes are characterized by an azimuthal wave covering the entire span of the cylinder with wave number equal to 2. The 5th mode of both the numerical and the experimental modes (here not shown for conciseness reasons) indeed exhibits the same structure, except for a rotation of $90°$ about the cylinder axis.

The first four POD modes related to the $Ro = 0.1$ computed from the numerical simulations and the experimental results are reported in figures 6.11-6.12 in a similar fashion to figures 6.9-6.10. In such a case, the similarities between the numerical

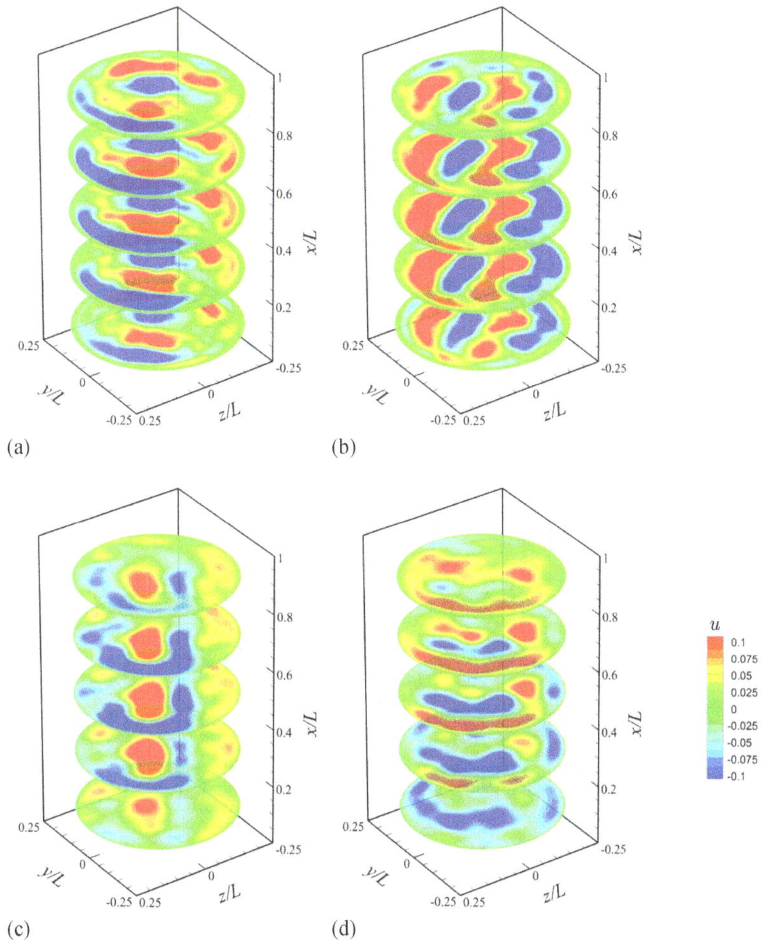

Figure 6.11: Comparison between the POD modes of the fluctuating velocity field for $Ro = 0.1$;. (a)-(b) 1st and 2nd POD modes from the numerical simulation at $Ra = 1.86 \times 10^8$ and $Pr = 7.6$; (c)-(d) 1st and 2nd POD modes from the experiment at $Ra = 2.86 \times 10^8$ and $Pr = 6.4$. Slices of the vertical velocity component at five different heights ($0.1L$, $0.3L$, $0.5L$, $0.7L$ and $0.9L$). Velocities are scaled by the maximum value of the velocity magnitude for each mode.

results and the experimental measurements are less pronounced. In particular, focusing on the 3rd and the 4th POD mode, it is possible to detect the presence of columnar structures characterized by opposite vertical velocities. It is evident that the structures observable in the numerical mode (figure 6.12a-b) are thinner and more numerous than those of the experimental case (figure 6.12c-d), in agreement with the analysis of the instantaneous snapshots (see figure 6.8).

Although pretty qualitative, the above discussion shows that, despite the net discrepancies between the time-mean behavior of the flow field, some dynamical features of the evolution of the rotating RB convection are similar between experiments

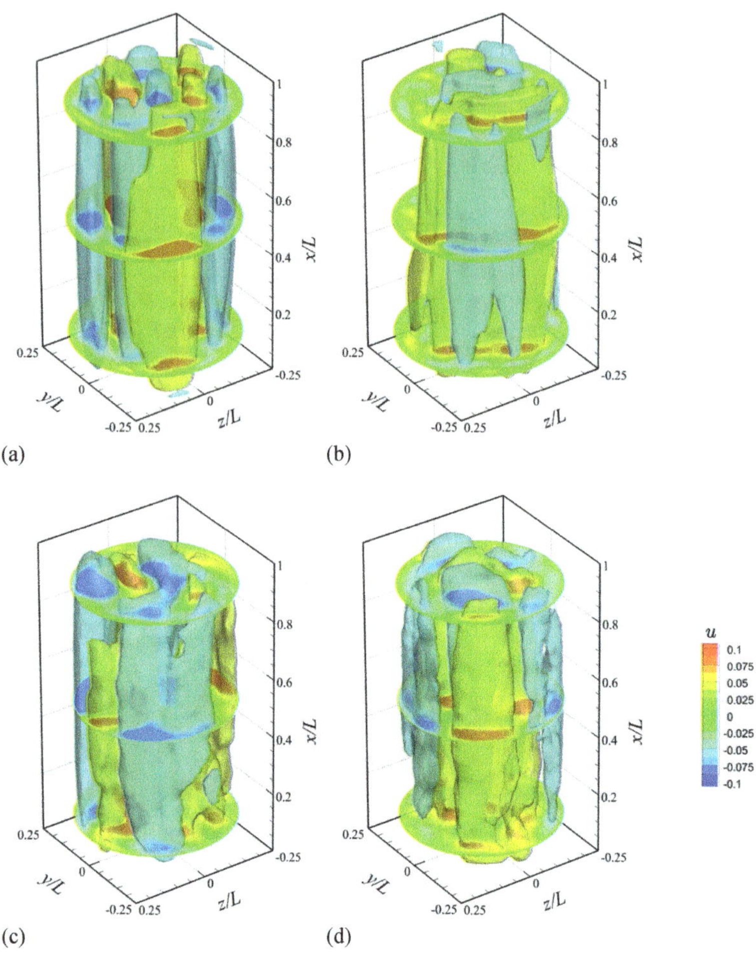

Figure 6.12: Comparison between the POD modes of the fluctuating velocity field for $Ro = 0.1$;. (a)-(b) 3rd and 4th POD modes from the numerical simulation at $Ra = 1.86 \times 10^8$ and $Pr = 7.6$; (c)-(d) 3rd and 4th POD modes from the experiment at $Ra = 2.86 \times 10^8$ and $Pr = 6.4$. Slices of the vertical velocity component at three different heights ($0.1L$, $0.5L$ and $0.9L$) and isosurfaces of positive and negative vertical velocity. Velocities are scaled by the maximum value of the velocity magnitude for each mode.

and simulations. The identification of the source of such discrepancies requires a deeper investigation from both the numerical and the experimental sides. Some plausible conjectures about this point are drawn in the next section.

# 6.3. Discussion about discrepancies between experiments and simulations

I N the previous section, the comparison between the experimental measurements and the numerical simulations of the RB convection in the $\Gamma = 0.5$ cylinder for $Ro = 0.25$ and $Ro = 0.1$ has shown weak agreement on the side of the time-averaged fields, whereas some similarities have been noticed in the analysis of the characteristic modes of the turbulent convection. The different values of $Ra$ and $Pr$ between the experiments and the simulations cannot justify these discrepancies. This statement is corroborated by the results of the numerical investigation of Kunnen *et al.* [71], which shows that rotating convection in a cylindrical sample with $\Gamma = 1$, $Ra = 1.00 \times 10^9$ and $Pr = 6.4$ exhibits a similar behavior to that observed in the present numerical investigation, in terms of the time-averaged evolution, as $Ro$ is decreased. Indeed, the influence of the aspect ratio on the unsteady dynamics is expected to be secondary only at high rotation rates, where the flow tends to two-dimensionalization due to the suppression of the vertical velocities and the vertical gradients of all the velocity components.

It is very difficult to argue about the origin of the discrepancies. The numerical code used (see section 3.2) has been validated by the comparison with several theoretical and experimental results as well as numerical simulations performed by different and independently-written codes. The non-rotating experiments presented and discussed in the current dissertation are themselves a further benchmark for the code and excellent agreement on this side has been found. On the other hand, the experiment/DNS mismatch might be related to subtle, unclear issues with both the experimental apparatus or the numerical setup. At this stage, we can only formulate suppositions that should be verified either experimentally or numerically.

First, we believe that one of the major discrepancies between the T-PIV experiments and the DNS is the mean distribution of the azimuthal velocities (compare figure 6.5c and 6.5f with figure 6.1k and 6.1o, respectively). The separation between the cyclonic and anticyclonic rotation in the bottom and the top half of the cylinder observed experimentally is an unexpected result, also in the light of the Taylor-Proudmann theorem which predicts the disappearance of the velocity vertical gradients in the bulk where viscosity is negligible. This could suggest that indeed the Coriolis forces are not dominant in the phenomenon under investigation. It is also known that, in addition to such forces, a uniform rotation introduces centrifugal forces, the importance of which with respect to the buoyancy forces is given by the Froude number (see equation (1.17)). In section 1.2, it has been mentioned that in canonical RB convection the centrifugal effects are typically neglected since $Fr \ll 1$; in the present case, indeed $Fr = 6.46 \times 10^{-3}$ for the larger rotation rate. It has already been noted that the definition of $Fr$ according to equation (1.17) (using the diameter as reference length) indeed determines the order of magnitude of the term related to the centrifugal forces with respect to the buoyancy forces. However, when rotation dominates, the centrifugal forces should be compared to the Coriolis ones rather than to buoyancy to determine if their effects are in fact negligible.

Referring to equation (1.12), the Coriolis force term is $-Ro^{-1}\,\widehat{\Omega} \times \mathbf{u}$, whereas the centrifugal term reads as $2Fr\Gamma^{-1}\theta\,\widehat{\mathbf{\Omega}}\times(\widehat{\mathbf{\Omega}}\times\mathbf{r})$. If we decrease the Rossby number by increasing the angular velocity, the Froude number correspondingly increases; however, for $\Omega < 1$ rad/s the increase in $Ro^{-1}$ is always greater than that in $Fr$. Nevertheless, it has to be considered that, with the free-fall scaling of the Oberbeck-Boussinesq equations, the term $Ro^{-1}$ does not represent the order of magnitude of the Coriolis forces since the non-dimensional velocity in the vector product $\widehat{\mathbf{\Omega}}\times\mathbf{u}$ is not unit order. Indeed, the centrifugal forces have to compete with the component of the Coriolis forces along the radial direction, which is given by $-Ro^{-1}\widehat{\mathbf{\Omega}}\times\mathbf{u}_\varphi$, being $\mathbf{u}_\varphi$ the non-dimensional azimuthal velocity component (vector). For the Rossby numbers investigated experimentally, we observe typical non-dimensional azimuthal velocities around $10^{-2}$–$10^{-3}$. This could lead to estimate the Coriolis forces to be the *same* or *one order of magnitude greater* than the centrifugal forces. Therefore, whether the centrifugal effects are negligible or not, it should be further investigated.

To the author's knowledge, few works have been devoted to the effects of the centrifugal forces on rotating turbulent convection. Among them, the work of Hart and Ohlsen [208] is worthy of citation. The latter was motivated by the intent of explaining the thermal offset observed in highly turbulent rotating convection experiments [209, 210]. As explained in the introduction of their paper, the loss of the Boussinesq symmetry (symmetry of the velocity field and anti-symmetry of the temperature field with respect to the mid-plane) cannot be explained by non-Boussinesq effects (which are surely negligible in the present investigation). Indeed, their mechanistic model associated the desymmetrization of the system to a mean meridional circulation induced by centrifugal buoyancy. The sketch of figure 4 of their article remarkably resembles our measurements of the velocity field in the case $Ro = 0.25$. A more recent publication on the subject is the paper by Horn and Aurnou [211]. This work interestingly shows that the transition to centrifugally dominated convection occurs for $Fr \gtrsim \Gamma$ and thus it is strictly dependent on the geometry of the convection cell. Based on such a result, the relevance of the centrifugal effects in the present case could be ruled out.

In other to assess the importance of the centrifugal forces in the experimental setup, direct numerical simulations including the centrifugal buoyancy term in the momentum balance equation have been performed for $Ra = 2.86 \times 10^8$, $Pr = 6.4$, $Ro = 0.25$ and $Fr = 4.14 \times 10^{-3}$, i.e. in the same conditions as the first experimental case of the current investigation. Such simulations have led to results coincident with those of the case $Fr = 0$ presented in section 6.1, thus they are not displayed here. These results confirm the expectation that the centrifugal forces do not play a role in the investigated configurations.

However, in order to show that the action of centrifugal forces causes an evolution similar to that observed experimentally, the results of numerical simulations performed in the case $Fr = 2\Gamma = 1$ are shown in figure 6.13. The numerical parameter settings of such simulations are essentially the same as those reported in Table 3.2, except $Ra = 2.86 \times 10^8$ and $Pr = 6.4$; moreover, the run duration is 600 non-dimensional time units and statistics have been collected over the last 400 times. The maps of the time- and azimuthally-averaged velocity components in figure 6.13 exhibit considerable similarities with the experimental measurements of figures

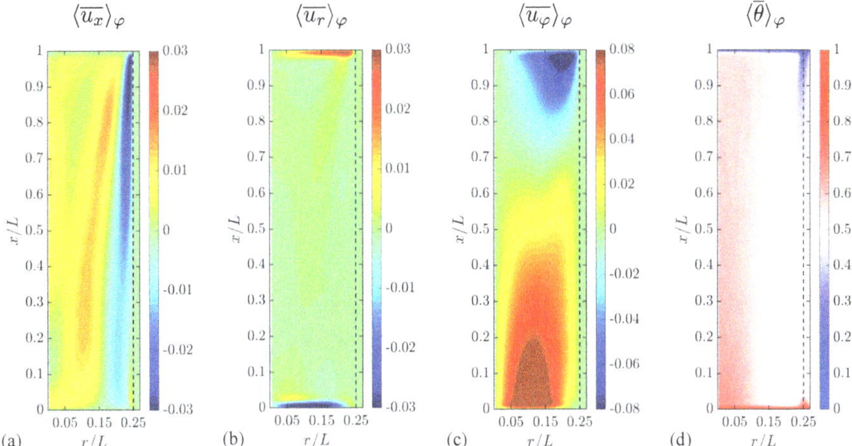


Figure 6.13: Effects of the centrifugal forces on rotating convection. Numerical simulations of the rotating convection for $Ra = 2.86 \times 10^8$, $Pr = 6.4$, $Ro = 0.25$ and $Fr = 1.0$. Temperature at the external side of the sidewall: $\theta_e = 0.5$. From left to right, time- and azimuthally-averaged fields of the vertical, radial and azimuthal velocity components and temperature. Velocities are scaled by the free-fall velocity, while temperature are relative to the temperatures of the bottom and scaled by the temperature difference across the fluid layer.

6.5a-c. In particular, it is still possible to detect a unique recirculation in the proximity of the sidewall that extends over the whole height of the cylindrical sample (figure 6.13a) associated with opposite azimuthal rotations in the bottom and the top half (figure 6.13c). Note that in such a configuration, differently from the experiment, cyclonic flow is observed near the bottom plate, while anticyclonic flow occurs in the top part of the cell; this is due to the fact that the background rotation is in the direction opposite to gravity ($\hat{\mathbf{g}}$ is parallel to the negative $x$-direction, while $\hat{\mathbf{\Omega}}$ is parallel to the positive $x$-direction). The time- and azimuthally-averaged temperature map (figure 6.13d) reveals a radial stratification of the flow with the cold, heavier fluid close to the sidewall and the hot, lighter fluid in the bulk, as expected result of the action of the centrifugal forces.

The above discussion points out that the experiments seem to be somehow affected by centrifugal buoyancy-like effects, although the occurrence of the latter is not ascribable to the extent of the background rotation by virtue of the small value of $Fr$ and the moderate temperature variations over the fluid domain. It should be also noticed that similar effects might be related to the influence of the temperature condition at the outside of the cylinder volume, which affects the interior by the heat transfer through the sidewall. In order to illustrate this, figure 6.14 reports the maps of the time- and azimuthally-averaged velocity components and temperature for the same working conditions as in figure 6.13 but a uniform temperature at the external side of the cylinder sidewall equal to $0.25$. Although, at a first glance, the structure of the mean flow field considerably resembles that of the experimental test and the simulation at unit Froude number, a more careful inspection discloses some relevant differences. More specifically, it is possible to observe the presence of a

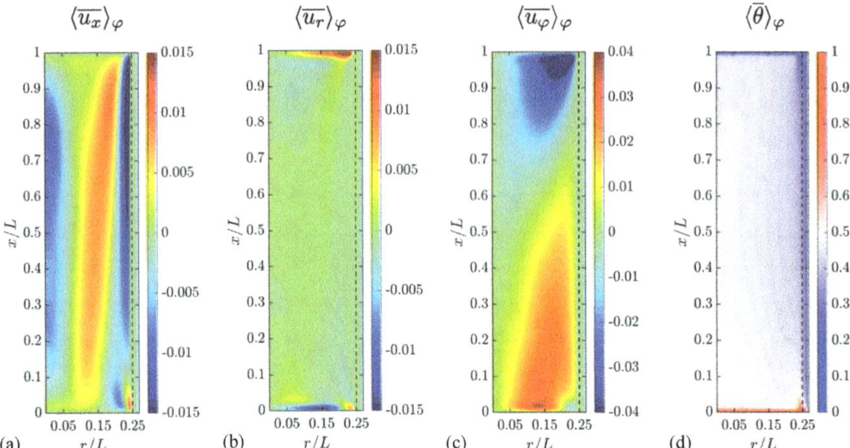


Figure 6.14: Effects of the temperature at the external side of the sidewall on rotating convection. Numerical simulations of the rotating convection for $Ra = 2.86 \times 10^8$, $Pr = 6.4$, $Ro = 0.25$ and $Fr = 4.14 \times 10^{-3}$. Temperature at the external side of the sidewall: $\theta_e = 0.25$. From left to right, time- and azimuthally-averaged fields of the vertical, radial and azimuthal velocity components. Velocities are scaled by the free-fall velocity, while temperature are relative to the temperatures of the bottom and scaled by the temperature difference across the fluid layer.

vertical descending current in the proximity of the cylinder axis, where in the above examined cases only small velocities are found; furthermore, the mid-plane symmetry of the temperature field is now broken and a considerable part of the convection cell is characterized by a time-average temperature lower than the average temperature between the top and the bottom of the cylinder (i.e., $\overline{\theta} < 0.5$).

In the current experiments, this influence of the external temperature condition is surely absent since the temperature of the tank is controlled via a cooling system driven by a TEC. As claimed in the introduction of this chapter, the tank temperature has been set to the average temperature between the bottom and the top temperature and maintained at this value with a stability better than $0.1\,°C$. The efficiency of the temperature control of the tank fluid has been assessed in a set of separate tests in both rotating and non-rotating conditions by measuring the local temperature simultaneously in different locations within the tank. Measurements have been performed with the cylinder in situ and setting the convection in the innerside; they show that the control based on the local measurement from a single thermal probe ensures an uniform temperature throughout the tank and thus, with a good confidence, on the entire external surface of the cylinder sidewall.

Numerical results in figure 6.14 suggest that any factor introducing a deviation of the average temperature over the convection cell from the average between the bottom and the top temperatures causes a desymmetrization of the flow field, leading to a vertical separation between cyclonic and anticyclonic flows. Among factors of this kind relevant to the present experimental apparatus are heating due to the absorption of the laser light from the sample water. If this effect is surely negligible in the case of pure water, the addition of particles could change the absorption coefficient with consequences on the heat transfer and ultimately fluid dynamics. However, it should be considered that the concentration of the seeding particles is indeed very small (the volumetric concentration is estimated to be lower than $0.001\%$ in the investigated conditions); moreover, if relevant, this effect should be evident also in the non-rotating case and, in any case, it would result in perceptible variations of the recorded light intensity within the images recorded by cameras in the direction of the cylinder axis.

Other motivations that could explain the experiment/DNS mismatch are the non-idealities of the experimental apparatus. Among them are the presence of the horizontal thermal gradients in the top and bottom plates, the non-uniform condition of the temperature at the external side of the cylinder (although the temperature control of the water tank ensures that the average temperature is equal to the average temperature over the fluid layer, there could be some spatial variations on this surface) and the influence of the particle deposits on the base plates, as well as the small fluctuations of the angular velocity during the rotation. The first three factors cited above are of course present also in the non-rotating experiment, thus it is hard to imagine that their effect changes significantly when a background rotation is applied to the system causing the desymmetrization of the flow field. It is plausible however that they have some influence on the specific characteristics of the columnar vortical structures. As concerns the stability of the rotational speed, it is very difficult to predict its effects since they are likely to depend on the particular residual vibrations and fluctuations of the rotating system. In our experiment, in the worst case, the standard deviation of the rotational speed is around $0.5\%$ of the average value. Although this is not at the cutting edge of the technology (most accurate systems reach stability around $0.01\%$), it is in line with other investigations on the rotating RBC and we are confident that the related effects are negligible at least on the time-average behavior.

## 6.4. Conclusions

IN this chapter, a numerical and experimental investigation of RB convection in a rotating reference frame has been carried out. The numerical simulations are performed at the same Rayleigh and Prandtl numbers of the non-rotating case ($Ra = 1.86 \times 10^8$ and $Pr = 7.6$) for several Rossby numbers varying from 0.05 to 1. In such simulations, the effects of the buoyancy centrifugal forces are neglected and rotation affects the flow field only through the action of the Coriolis forces. The behaviors of both the flow field and the global heat transfer have been analyzed numerically, showing a very good agreement with results presented in literature. Experimental measurements at slightly different Rayleigh and Prandtl numbers ($Ra = 2.86 \times 10^8$ and $Pr = 6.4$) and for $Ro = 0.25$ and $Ro = 0.1$ have been presented and discussed in comparison with the numerical results. A substantial mismatch between experiments and DNS has been observed, which could be related, on one side, to flaws of the physical model underlying numerical simulations, on the other, to non-idealities of the experimental apparatus that makes the experimentally reproduced condition not compliant with the canonical phenomenon.

The points of major concern in the above regards are:

- in the numerical simulations, the increase in $Ro$ results in the confinement of the radial recirculations observed in the non-rotating mean flow field to a thin region adjacent to the sidewall (Stewartson boundary layer), where the flow is subjected to a cyclonic rotation opposite to the anticyclonic motion in the middle part of the cylindrical sample. Conversely, experimental fields exhibit a breakdown of the up/down symmetry with the occurrence of a unique recirculation close to the sidewall and extending through the whole cylinder height and an axial separation of oppositely rotating currents;

- vertically aligned vortical structures, resembling Taylor columns, are detected both in numerical simulations and experiments, but in the latter case lower concentration and greater characteristic sizes are observed;

- corresponding POD modes exhibit an almost identical structure between experiments and simulations, suggesting that, despite the significant differences in the mean behavior, flow fields observed numerically and experimentally share somehow similar features;

- experimental measurements suggest the occurrence of centrifugal buoyancy-like effects in thermal convection. Several could be the factors originating such effects:

  - relevance of the centrifugal forces;

  - influence of the temperature condition at the external side of the cylinder sidewall, or better deviation of the tank bulk temperature from the average temperature over the convection cell;

  - increase of the average temperature over the convective layer due to fluid internal heating related to absorption of laser energy.

Specific numerical simulations have ruled out the first effect, demonstrating the negligibility of the buoyancy centrifugal forces for the present values of $Fr$. On the other side, the tank temperature was controlled during the experiments to avoid the second problem, while absorption of laser energy by water should be very small despite the seeding particles could affect the absorption coefficient. Therefore, no persuasive arguments are in fact identified to explain this behavior clearly observed in the experiments. It remains an open issue.

   CHAPTER 6. Investigation of rotating Rayleigh-Bénard convection

**7**

# Conclusions and perspectives

THIS thesis has addressed the application of time-resolved tomographic particle image velocimetry to the study of Rayleigh-Bénard convection without and with background rotation. In addition, direct numerical simulations have been carried out and compared to the 3D experimental measurements. The complementary use of computational fluid dynamics and experiments has made it possible on one side to validate the correctness and completeness of the numerical models, on the other side, to elucidate the parameters and the boundary conditions that unavoidably affect the experiment and cause it to deviate from the canonical phenomenon.

The first goal of this work has been the development of an experimental apparatus suitable for the optical investigation of the RB convection in the whole interior of a cylinder with aspect ratio equal to one half. The requirement of optical accessibility prevents the possibility of an adequate thermal insulation of the cylinder sidewall. As a consequence, the thermal condition of the cylinder sidewall has a significant influence on the flow dynamics. This is confirmed by our experiments in non-rotating conditions. In order to control such an effect, an appropriate temperature control system has been designed to maintain the temperature at the external side of the cylinder sidewall constant to a predefined value (in the present case the average temperature between the bottom and the top plate). The presence of the cylinder sidewall with its physical properties has been included in the simulation to avoid discrepancies on that side; the comparison with the experimental measurements in the non-rotating case shows that the code describes correctly such an influence.

The detrimental effects of the optical distortion caused by the cylinder sidewall on the tomographic reconstruction have been largely discussed and experimentally assessed. In order to correctly model such a distortion an innovative camera calibration model has been developed. This model consists essentially in a modification of the pinhole camera model by means of a refraction correction for the cylindrical distortion based on Snell's law. Relying upon physical laws, the innovative model makes use of a relatively small number of parameters, which have a clear geometrical and physical meaning (and thus are prone to easy check). An additional advantage of the model is the possibility of performing the camera calibration without the need of sweeping a calibration target through the cylinder interior; in this regard, a specific calibration procedure has been defined and exploited in the present experimental investigation. The innovative model has been comparatively assessed against classical camera models, like the classical pinhole camera model (proven to be inadequate for imaging of the whole cylinder) and the polynomial functions. The superior performance of the model has been shown in terms of both the residual errors of calibration and the number of employed parameters.

An extensive study of the RB convection in a cylinder with $\Gamma = 1/2$ in non-rotating frame at $Ra = 1.86 \times 10^8$ and $Pr = 7.6$ has been carried out both experimentally and numerically. A good agreement between the experimental measurements and the numerical results has been found in the analysis of both the time-averaged velocity fields and the statistical behavior of the LSC. The mean velocity field has been found to exhibit axisymmetric structure in agreement with previous numerical and experimental investigations; this structure has been related to the quite complex dynamics of the LSC and plumes. The characteristic modes of the turbulent convection have been extracted by modal decomposition of the fluctuating velocity field. The first POD modes capture the principal states of the LSC that have been observed in previous numerical and experimental studies, but never, to the author's knowledge, in a 3D experimental investigation; also some of the oscillatory modes of the LSC are captured by the POD modes. Such results can be considered as a visual, concrete representation of the flow modes of the turbulent convection detected in previous experimental investigation with only 2D optical techniques or different tools, like the multi-thermal probe approach. The POD modes determined from the numerical simulations are essentially the same as in the experimental case, except for the emergence of two additional modes that can be associated with the torsional oscillation of the LSC. Of much more interest is that the first two POD modes of the temperature fluctuation are not representative of the single-roll state of the LSC nor the double-roll one; indeed, undefined states which might be associated with a transition between the states of the LSC are observed.

The statistical occurrence and behavior of the LSC states has been also studied. As a first step, methods consolidated in literature have been used for this purpose; the obtained results show an excellent agreement with the experimental investigations of Xi and Xia [58] and Weiss and Ahlers [63] and the numerical study of Stevens *et al.* [101], although our experiment is limited to a considerably lower duration (four hours). Subsequently, the relationship of the SRS and the DRS with the low order POD modes has been investigated and a new method to identify the LSC based on a POD analysis has been presented. Such a method substantially agrees with the techniques used in literature, however a bias in the identification of the LSC orientation related to the latter has been found. We believe that the POD-based method is more accurate and robust, although it involves the measurements of the entire flow field, which could be impracticable for very long times.

As concerns the investigation of the rotating RB convection, numerical results have been presented to show the influence of the rotation on the mean flow field and the global heat transfer. In this regard, we have clarified the role of the Ekman and the Stewartson boundary layers as the Rossby number is decreased and highlighted the progressive suppression of the vertical and radial velocities and the establishment of opposite cyclonic and anticyclonic rotation in the bulk and near the sidewall of the cylindrical sample. The comparison of such numerical results with the experimental measurements has not been as successful as in the non-rotating case. Experimental measurements have been performed for $Ro = 0.25$ and $Ro = 0.1$ and in both cases a substantial disagreement between the time-averaged velocity fields computed from simulations and measured in the experiments has been found. Conversely, the characteristic modes identified by POD show some similarities between DNS and experiments. It is very hard to argue about the origin of such discrepancies, since they could arise either from a lack of completeness of the numerical model or a specific property of the experimental apparatus that changes the dynamics of the phenomenon with respect to the canonical case. It has been noted that centrifugal buoyancy-like effects occur in the experimental case. With the aid of numerical simulations, it has been demonstrated that these effects are not ascribable to the action of centrifugal buoyancy forces by virtue of the smallness of the Froude number. Furthermore, it has been noted that a similar influence might be due to a difference between the average temperature over the convective layer and the temperature outside of the cylinder, which affects the flow motion by heat transfer through the sidewall. However, in the rotating experiments, the tank temperature has been accurately controlled and thus such an effect can be ruled out. Other possible explanations have been considered, however none of them seems to be conclusive.

The unresolved discrepancy between the experiments and the simulations of the rotating convection leaves much room for future work. A possible route in this regard could be to investigate a wider range of operating conditions both experimentally and numerically and assess if the observed discrepancies persist also in very different conditions. For instance, it could be interesting to study the regime of very slow angular velocities where the centrifugal forces are expected to be in any case negligible. Currently, this is not possible with our rotating system since high stability of the rotation is not ensured in such a regime.

# Proper orthogonal decomposition

T HE proper orthogonal decomposition (POD) is a statistical method for extracting a basis for modal decomposition from an ensemble of signals [196]. POD provides a linear decomposition of the available signals, which, among all the linear decompositions, is optimal from an energetic standpoint, as explained in the following. Such a feature has made it become a fundamental tool in the analysis of complex turbulent flows, often preferable to other techniques due to the simple and clear interpretation to which POD modes are prone. In fact, POD generally allows for identification of the most energetic coherent structures, which play a fundamental role in several processes like mixing, heat transfer, generation of aerodynamic forces, etc..

The first application of POD to the investigation of turbulent flows dates back to Lumley [212]. However, this technique was independently formulated by different scientists some years before [213–217] and, in other contexts, it is also known as Karhunen–Loève decomposition or principal component analysis.

The mathematical foundation of POD is provided by the spectral theory of compact self-adjoint operators, which applies to both discrete and continuous signals. A general presentation of such a theory goes beyond the scope of the present dissertation. Since both experimental measurements and numerical results are obtained in a discrete form, in the following we only deal with POD of discrete signals, explaining its relationship with single-value decomposition (SVD). An introduction to extended POD is also given in the final part of this appendix.

The following treatment is inspired by [218–220] for POD and by [200–202] for extended POD.

## A.1. Mathematical framework

L ET us suppose to collect a set of $n$ observations, with each observation $\mathbf{v}_j$ consisting of a vector of $m$ elements $v_{ij}$, i.e. $\mathbf{v}_j \in \mathbb{R}^m$. These elements might be, for instance, the measurements of the same flow quantity (velocity, temperature, etc.)

in different locations of the investigated volume. The observations are arranged in a $m \times n$ matrix $\mathbf{V} = [\mathbf{v}_1, \mathbf{v}_2, \ldots, \mathbf{v}_n]$, called the observation matrix.

We are interested in determining the vectors $\psi$ that best approximate all the observations $\mathbf{v}_j$ in an average sense. Mathematically, this problem can be expressed as:

$$\max_{\psi \in \mathbb{R}^m} \sum_{j=1}^{n} \frac{|(\mathbf{v}_j, \psi)|^2}{|(\psi, \psi)|} \tag{A.1}$$

where the scalar product between the observation $\mathbf{v}_j$ and the vector $\psi$ is defined by:

$$(\mathbf{v}_j, \psi) = \mathbf{v}_j^{\mathrm{T}} \psi = \sum_{i=1}^{m} v_{ij} \psi_i. \tag{A.2}$$

Note that the problem (A.1) is equivalent to the problem:

$$\max_{\psi \in \mathbb{R}^m} \sum_{j=1}^{n} |(\mathbf{v}_j, \psi)|^2 \tag{A.3}$$

with the constraint that the norm of $\psi$ is unit:

$$||\psi||^2 = |(\psi, \psi)| = 1. \tag{A.4}$$

The latter optimization problem can be solved using the method of Lagrange multipliers. Accordingly, we can define the Lagrange functional:

$$\mathcal{F}(\psi, \lambda) = \sum_{j=1}^{n} |(\mathbf{v}_j, \psi)|^2 + \lambda(1 - ||\psi||^2) \tag{A.5}$$

and impose the necessary optimality condition for the existence of a solution $\psi$:

$$\nabla \mathcal{F}(\psi, \lambda) = 0. \tag{A.6}$$

After some algebra, this finally yields:

$$\mathbf{V}\mathbf{V}^{\mathrm{T}}\psi = \lambda\psi. \tag{A.7}$$

Equation (A.7) is an eigenvalue problem in $\mathbb{R}^m$. Being the matrix $\mathbf{V}\mathbf{V}^{\mathrm{T}}$ positive semi-definite, it has $m$ nonnegative eigenvalues $\lambda_1 \geq \lambda_2 \geq \cdots \geq \lambda_m \geq 0$ and the corresponding eigenvectors $\psi_1, \psi_2, \ldots, \psi_m$ can be chosen such that they are pairwise orthogonal. These eigenvectors are the POD modes, while, from the same equation (A.7), the eigenvalues $\lambda_j$ with $j = 1, 2, \ldots, m$ can be easily regarded as the contribution of these modes to the average energy budget of the observations $\sum_j (\mathbf{v}_j, \mathbf{v}_j)$. In fact, by left-multiplying each side of equation (A.7) by the vector $\psi^{\mathrm{T}}$ and considering that $\psi^{\mathrm{T}}\psi = ||\psi||^2 = 1$, it follows that:

$$\psi^{\mathrm{T}}\mathbf{V}\mathbf{V}^{\mathrm{T}}\psi = (\mathbf{V}^{\mathrm{T}}\psi)^{\mathrm{T}}(\mathbf{V}^{\mathrm{T}}\psi) = ||\mathbf{V}^{\mathrm{T}}\psi||^2 = \sum_{j=1}^{m} |(\mathbf{v}_j, \psi)|^2 = \lambda. \tag{A.8}$$

The latter equation also proves that the POD eigenvalues are nonnegative. The set of such eigenvalues, ordered in a decreasing way, is the so-called *POD spectrum*.

Since the POD modes form an orthonormal basis of $\mathbb{R}^m$, i.e. $\boldsymbol{\psi}_l^{\mathrm{T}}\boldsymbol{\psi}_m = \delta_{lm}$ with $\delta_{lm}$ being the Kronecker delta, each column of $\mathbf{V}$, i.e. each observation $\mathbf{v}_j$, can be regarded as a linear combination of these modes:

$$\mathbf{v}_j = \sum_{k=1}^{m} \beta_{jk}\boldsymbol{\psi}_k. \tag{A.9}$$

where the coefficients $\beta_{jk}$ are indeed the projection of the observations on the same POD modes, as proven by the following relationship:

$$\mathbf{v}_j^{\mathrm{T}}\boldsymbol{\psi}_k = \left(\sum_{l=1}^{m} \beta_{jl}\boldsymbol{\psi}_l^{\mathrm{T}}\right)\boldsymbol{\psi}_k = \sum_{l=1}^{m} \beta_{jl}(\boldsymbol{\psi}_k^{\mathrm{T}}\boldsymbol{\psi}_k) = \sum_{l=1}^{m} \beta_{jl}\delta_{lk} = \beta_{jk}. \tag{A.10}$$

Equation (A.9) constitutes the modal decomposition of the observations $\mathbf{v}_j$ on the basis given by the POD modes, often referred also as spectral decomposition.

As previously mentioned, the POD is an optimal decomposition in an energetic sense. The following proposition establishes such a property of optimality of POD in a more formal fashion.

**Proposition** *Let* $\mathbf{V} = [\mathbf{v}_1, \mathbf{v}_2, \ldots, \mathbf{v}_n]$ *be a set of* $n$ *observations* $\mathbf{v}_j \in \mathbb{R}^m$. *Let*

$$\mathbf{v}_j = \sum_{k=1}^{m} \beta_{jk}\boldsymbol{\psi}_k$$

*be the POD of such a data set and let* $\boldsymbol{\varphi}_k \in \mathbb{R}^m$ *be an arbitrary orthonormal basis set such that:*

$$\mathbf{v}_j = \sum_{k=1}^{m} \gamma_{jk}\boldsymbol{\varphi}_k.$$

*Then the following holds:*

$$\sum_{j=1}^{m} \left\| \mathbf{v}_j - \sum_{k=1}^{p} \beta_{jk}\boldsymbol{\psi}_k \right\|^2 \leq \sum_{j=1}^{m} \left\| \mathbf{v}_j - \sum_{k=1}^{p} \gamma_{jk}\boldsymbol{\varphi}_k \right\|^2$$

*for any* $p$ *such that* $1 \leq p \leq m$.

The demonstration of this proposition is not presented here. For further details, the reader is referred to classical textbooks like [218].

## A.2.  Relationship with SVD and method of snapshots

Hereinbefore, the POD modes have been defined as the solutions of the eigenvalue problem (A.7). It is noted that since the temporal covariance matrix of the observations is defined as $\mathbf{C_V} = 1/(n-1)\mathbf{VV}^\mathrm{T}$, the POD modes are in fact the eigenvectors of such a matrix. In the following, we show the relationship of the POD modes with the SVD of the observation matrix.

The theorem concerning SVD ensures that any matrix $\mathbf{V} \in \mathbb{C}^{m \times n}$ can be always decomposed in the product of three matrices as follows:

$$\mathbf{V} = \mathbf{\Psi}\mathbf{\Sigma}\mathbf{\Phi}^* \tag{A.11}$$

where:

- $\mathbf{\Psi} = [\psi_1, \psi_2, \ldots, \psi_m] \in \mathbb{C}^{m \times m}$ is a unitary matrix, i.e. it satisfies the condition $\mathbf{\Psi}\mathbf{\Psi}^* = \mathbf{\Psi}^*\mathbf{\Psi} = \mathbf{I}_m$ with $\mathbf{I}_m$ the $m \times m$ identity matrix; the columns $\psi_j$ with $j = 1, 2, \ldots, m$ are called *left-singular vectors*;

- $\mathbf{\Sigma} \in \mathbb{R}^{m \times n}$ is a diagonal matrix, the diagonal element of which are nonnegative real numbers, called *singular values* of the matrix $\mathbf{V}$, and are ordered in a decreasing way: $\sigma_1 \geq \sigma_2 \geq \cdots \geq \sigma_{\min(m,n)} \geq 0$;

- $\mathbf{\Phi} = [\phi_1, \phi_2, \ldots, \phi_n] \in \mathbb{C}^{n \times n}$ is a unitary matrix, i.e. it satisfies the condition $\mathbf{\Phi}\mathbf{\Phi}^* = \mathbf{\Phi}^*\mathbf{\Phi} = \mathbf{I}_n$ with $\mathbf{I}_n$ the $n \times n$ identity matrix; the columns $\phi_i$ with $i = 1, 2, \ldots, n$ are called *right-singular vectors*.

In the above equations, the asterisk denotes the conjugate transpose operator. When this operator is applied to real-valued matrices, it coincides simply with the transpose operator. It could be demonstrated that if $\mathbf{V}$ is real-valued (which is the case of an observation matrix comprising experimental measurements), also the matrices $\mathbf{\Psi}$ and $\mathbf{\Phi}$ are real-valued. The following discussion focuses on the latter case, therefore the symbol $*$ is always replaced with $^\mathrm{T}$ without any disclaimer.

By using the properties of the unitary matrices, the transpose of the observation matrix is given by $\mathbf{V}^\mathrm{T} = \mathbf{\Phi}\mathbf{\Sigma}^\mathrm{T}\mathbf{\Psi}^\mathrm{T}$ and finally it is possible to obtain that:

$$\mathbf{VV}^\mathrm{T} = \mathbf{\Psi}\mathbf{\Sigma}\mathbf{\Phi}^\mathrm{T}\mathbf{\Phi}\mathbf{\Sigma}^\mathrm{T}\mathbf{\Psi}^\mathrm{T} = \mathbf{\Psi}\mathbf{\Sigma}\mathbf{I}_n\mathbf{\Sigma}^\mathrm{T}\mathbf{\Psi}^\mathrm{T} = \mathbf{\Psi}\mathbf{\Sigma}\mathbf{\Sigma}^\mathrm{T}\mathbf{\Psi}^\mathrm{T} \tag{A.12}$$

which can be also written as:

$$\mathbf{VV}^\mathrm{T}\mathbf{\Psi} = \mathbf{\Psi}\mathbf{\Sigma}\mathbf{\Sigma}^\mathrm{T} \quad \text{or} \quad \mathbf{VV}^\mathrm{T}\psi_j = \sigma_j^2\psi_j \text{ with } j = 1, 2, \ldots, m. \tag{A.13}$$

Equation (A.13) proves that the POD modes coincide with the left-singular vectors of the observation matrix, while the POD eigenvalues are equal to the squared singular values.

It is also interesting to note that the POD modes can be derived from the eigenvectors of the spatial covariance matrix defined as $\mathbf{C'_V} = 1/(m-1)\mathbf{V}^\mathrm{T}\mathbf{V}$, which coincide with the right-singular vectors. In fact, analogously to equation (A.12):

$$\mathbf{V}^\mathrm{T}\mathbf{V} = \mathbf{\Phi}\mathbf{\Sigma}^\mathrm{T}\mathbf{\Psi}^\mathrm{T}\mathbf{\Psi}\mathbf{\Sigma}\mathbf{\Phi}^\mathrm{T} = \mathbf{\Phi}\mathbf{\Sigma}^\mathrm{T}\mathbf{I}_m\mathbf{\Sigma}\mathbf{\Phi}^\mathrm{T} = \mathbf{\Phi}\mathbf{\Sigma}^\mathrm{T}\mathbf{\Sigma}\mathbf{\Phi}^\mathrm{T} \tag{A.14}$$

or, in a form similar to equation (A.13),

$$\mathbf{V}^{\mathrm{T}}\mathbf{V}\boldsymbol{\Phi} = \boldsymbol{\Phi}\boldsymbol{\Sigma}^{\mathrm{T}}\boldsymbol{\Sigma} \quad \text{or} \quad \mathbf{V}^{\mathrm{T}}\mathbf{V}\boldsymbol{\phi}_i = \sigma_i^2 \boldsymbol{\phi}_i \text{ with } i = 1, 2, \ldots, n. \tag{A.15}$$

It should be noted that equation (A.15) constitutes the solution of an eigenvalue problem in $\mathbb{R}^n$. In those applications where measurements of the same quantity in a large number of spatial locations are performed at a relatively small number of different time instants (as typically in T-PIV applications), it occurs that $m \gg n$ and solving the problem underlying equation (A.15) is significantly cheaper than solving the problem given by equation (A.7). When adopting such a strategy, however, the temporal modes $\boldsymbol{\phi}_i$ are obtained instead of the spatial POD modes $\boldsymbol{\psi}_j$. The latter can be subsequently determined by inverting equation (A.11) as follows:

$$\boldsymbol{\Psi} = \mathbf{V}\boldsymbol{\Phi}\boldsymbol{\Sigma}^{+} \tag{A.16}$$

where $\boldsymbol{\Sigma}^{+}$ is the pseudo-inverse matrix of the diagonal matrix $\boldsymbol{\Sigma}$. This method is also known as "method of snapshots" and was first proposed by Sirovich [221].


## A.3. Extended POD

E XTENDED POD [200–202] is a technique used in the analysis of the correlation between two sets of observations, possibly related to different quantities, which allows to identify the characteristic modes of the part of the first data set that correlates to the second data set. In the following, the mathematical framework of such a technique is briefly introduced.

Let $\mathbf{A} = [\mathbf{a}_1, \mathbf{a}_2, \ldots, \mathbf{a}_n] \in \mathbb{R}^{m_A \times n}$ and $\mathbf{B} = [\mathbf{b}_1, \mathbf{b}_2, \ldots, \mathbf{b}_n] \in \mathbb{R}^{m_B \times n}$ be two different sets of observations correlated with each other. Without loss of generality, we suppose that such observations have been collected at the same times, although they might correspond to different locations of the investigated volume. For instance, the observations $\mathbf{a}_j$ might be the velocity measurements of the flow in the whole flow field under investigation, whereas $\mathbf{b}_j$ might be the temperature measurements taken simultaneously only in a portion of it. From equation (A.16) the POD modes of the first data set can be obtained as $\boldsymbol{\Psi}_{\mathbf{A}} = \mathbf{A}\boldsymbol{\Phi}_{\mathbf{A}}\boldsymbol{\Sigma}_{\mathbf{A}}^{+}$, i.e. as a weighted average of the observations $\mathbf{a}_j$ with weights given by the elements of the matrix $\boldsymbol{\Phi}_{\mathbf{A}}\boldsymbol{\Sigma}_{\mathbf{A}}^{+}$. The extended POD modes of the ensemble of observations $\mathbf{B}$ are defined as:

$$\overline{\boldsymbol{\Psi}}_{\mathbf{B}}^{(\mathrm{ext})} = \mathbf{B}\boldsymbol{\Phi}_{\mathbf{A}}\boldsymbol{\Sigma}_{\mathbf{A}}^{+}, \tag{A.17}$$

therefore they are obtained by averaging the observations $\mathbf{b}_j$ with the same weights used to obtain the modes $\boldsymbol{\Psi}_{\mathbf{A}}$ from the observations $\mathbf{a}_j$. It is worth noting that the extend POD modes provided by equation (A.17) are not normalized and a more general definition would be:

$$\boldsymbol{\Sigma}_{\mathbf{B}}^{(\mathrm{ext})}\boldsymbol{\Psi}_{\mathbf{B}}^{(\mathrm{ext})} = \mathbf{B}\boldsymbol{\Phi}_{\mathbf{A}}\boldsymbol{\Sigma}_{\mathbf{A}}^{+} \tag{A.18}$$

with $\boldsymbol{\Psi}_{\mathbf{B}}^{(\mathrm{ext})}$ being an orthonormal matrix.

Following the above definition of the extended POD modes, the observations $\mathbf{b}_j$ can be decomposed as follows:

$$\mathbf{b}_j = \mathbf{b}_j^{(C)} + \mathbf{b}_j^{(D)} = \sum_{k=1}^{m_B} \beta_{jk}^{(C)} \psi_{\mathbf{B}}^{(\text{ext})} + \mathbf{b}_j^{(D)} \tag{A.19}$$

where $\mathbf{b}_j^{(C)}$ is the correlated part of the observations, while $\mathbf{b}_j^{(D)}$ is the decorrelated one. Equation (A.19) corresponds to an analogous decomposition of the observation matrix: $\mathbf{B} = \mathbf{B}^{(C)} + \mathbf{B}^{(D)}$. It is possible to demonstrate that the matrix $\mathbf{B}^{(C)} = [\mathbf{b}_1^{(C)}, \mathbf{b}_2^{(C)} m, \ldots, \mathbf{b}_n^{(C)}]$ is the only part of $\mathbf{B}$ that correlates with the matrix $\mathbf{A}$, i.e., $\mathbf{A}^{\mathsf{T}} \mathbf{B}^{(D)} = 0$. The proof of such a statement is not provided here for brevity. Further information is available in [201].

# References


[1] H. Bénard, *Les tourbillons cellulaires dans une nappe liquide,* Rev. Gen. Sci. Pures Appl. **11**, 1261 (1900). [page 1].

[2] H. Bénard, *Les tourbillons cellulaire dans nappe liquide transportant de la chaleur purconvections en regime permanent,* Rev. Gen. Sci. Pures Appl. Bull. Assoc. **11**, 1309 (1900). [page 1].

[3] E. H. Weber, *Mikroskopische Beobachtungen sehr gesetzmässiger Bewegungen, welche die Bildung von Niederschlägen harziger Körper aus Weingeist begleiten,* Ann. Phys. **170**, 447 (1855). [page 1].

[4] J. Thompson, *On a changing tessellated structure in certain fluid,* Proc. R. Philos. Soc. Glasg. **13**, 464 (1882). [page 1].

[5] L. Rayleigh, *On convection currents in a horizontal layer of fluid, when the higher temperature is on the under side,* Lond. Edinb. Dubl. Phil. Mag. **32**, 529 (1916). [page 1].

[6] H. Jeffreys, *The stability of a layer of fluid heated below,* Lond. Edinb. Dubl. Phil. Mag. **2**, 833 (1926). [page 1].

[7] H. Jeffreys, *Some cases of instability in fluid motion,* Proc. R. Soc. Lond. A **118**, 195 (1928). [page 1].

[8] A. R. Low, *On the criterion for stability of a layer of viscous fluid heated from below,* Proc. R. Soc. Lond. A **125**, 180 (1929). [page 1].

[9] H. Bénard, *Sur les tourbillons cellulaires, les tourbillons enbandes, et la théorie de Rayleigh,* Soc. Frang. Phys. **266**, 112 (1928). [page 1].

[10] H. Bénard, *Sur la limite théorique de stabilité de l'équilibre préconvectif de Rayleigh, comparée à celle que donne l'étude expérimentale des tourbillons cellulaires de Bénard,* in *Proceedings of the 4th International Congres of Applied Mechanics, Cambridge, 1934; Journées de Mécanique des Fluides de l'Université de Lille II, Ed. 317, 1935* (Chiron, 1935). [page 1].

[11] M. J. Block, *Surface tension as the cause of Bénard cells and surface deformation in a liquid film,* Nature **178**, 650 (1956). [page 2].

[12] J. R. A. Pearson, *On convection cells induced by surface tension,* J. Fluid Mech. **4**, 489 (1958). [page 2].

[13] H. Bénard, *Les tourbillons cellulaires dans une nappe liquide.-Méthodes optiques d'observation et d'enregistrement,* J. Phys. Theor. Appl. **10**, 254 (1901). [page 2].

[14] S. Rahmstorf, *Thermohaline ocean circulation,* in *Encyclopedia of Quaternary Sciences,* edited by S. A. Elias (Elsevier, Amsterdam, 2006) pp. 739–750. [page 2].

[15] Hinode JAXA/NASA, *Developing Sunspot,* `https://www.nasa.gov/mission_pages/hinode/solar_022.html` (2007), accessed: 2018-08-20. [page 2].

[16] NASA/JPL/Space Science Institute, *Map of Jupiter's South*, `https://www.nasa.gov/multimedia/imagegallery/image_feature_539.html` (2007), accessed: 2018-08-20. [page 2].

[17] S. Chandrashekhar, *Hydrodynamic and Hydromagnetic Stability* (Dover, New York, 1981). [page 3].

[18] P. G. Drazin and W. H. Reid, *Hydrodynamic stability* (Cambridge University Press, Cambridge, England, 1981). [page 3].

[19] G. Hadley, *Concerning the cause of the general trade-winds*, Philos. Trans. Royal Soc. **39**, 58 (1735). [page 3].

[20] S. Rahmstorf, *The thermohaline ocean circulation: A system with dangerous thresholds?* Clim. Change **46**, 247 (2000). [page 3].

[21] F. M. Richter, *Mantle convection models*, Annu. Rev. Earth Planet. Sci. **6**, 9 (1978). [page 3].

[22] W. J. Morgan, *Deep mantle convection plumes and plate motions*, Am. Assoc. Pet. Geol. Bull. **56**, 203 (1972). [page 3].

[23] G. A. Glatzmaier and P. H. Roberts, *Dynamo theory then and now*, Int. J. Eng. Sci. **36**, 1325 (1998). [page 3].

[24] G. A. Glatzmaier and P. H. Roberts, *A three-dimensional convective dynamo solution with rotating and finitely conducting inner core and mantle*, Phys. Earth Planet. Inter. **91**, 63 (1995). [pages 3 and 12].

[25] H. C. Spruit, A. Nordlund, and A. M. Title, *Solar convection*, Annu. Rev. Astron. Astrophys. **28**, 263 (1990). [page 3].

[26] M. S. Miesch, *The coupling of solar convection and rotation*, in *Helioseismic Diagnostics of Solar Convection and Activity* (Springer, 2000) pp. 59–89. [page 3].

[27] F. H. Busse, *Convection driven zonal flows and vortices in the major planets*, Chaos **4**, 123 (1994). [page 3].

[28] J. P. Johnston, *Effects of system rotation on turbulence structure: a review relevant to turbomachinery flows*, Int. J. Rotating Mach. **4**, 97 (1998). [page 3].

[29] R. van Wissen, M. Golombok, and J. J. H. Brouwers, *Gas centrifugation with wall condensation*, AlChE J. **52**, 1271 (2006). [page 3].

[30] R. D. Oldham, *The constitution of the interior of the Earth, as revealed by earthquakes*, Quart. J. Geol. Soc. **62**, 456 (1906). [page 3].

[31] K.-Q. Xia, *Current trends and future directions in turbulent thermal convection*, Theor. Appl. Mech. Lett. **3** (2013), 10.1063/2.1305201. [page 4].

[32] G. Ahlers, S. Grossmann, and D. Lohse, *Heat transfer and large scale dynamics in turbulent Rayleigh-Bénard convection*, Rev. Mod. Phys. **81**, 503 (2009). [pages 4 and 8].

[33] D. Lohse and K.-Q. Xia, *Small-scale properties of turbulent Rayleigh-Bénard convection*, Annu. Rev. Fluid Mech. **42**, 335 (2010). [page 4].

[34] R. J. Stevens, H. J. Clercx, and D. Lohse, *Heat transport and flow structure in rotating Rayleigh–Bénard convection*, Eur. J. Mech. B/Fluids **40**, 41 (2013). [pages 4 and 15].

[35] A. Oberbeck, *Über die wärmeleitung der flüssigkeiten bei berücksichtigung der strömungen infolge von temperaturdifferenzen,* Ann. Phys. **243**, 271 (1879). [page 4].

[36] J. Boussinesq, *Théorie analytique de la chaleur: mise en harmonie avec la thermodynamique et avec la théorie mécanique de la lumière,* Vol. 2 (Gauthier-Villars, 1903).

[37] L. D. Landau and E. M. Lifshitz, *Fluid Mechanics* (Pergamon Press, Oxford, 1987). [page 4].

[38] H. P. Greenspan, *The theory of rotating flows* (Breukelen Press, Brookline, 1990). [page 8].

[39] P. K. Kundu, I. M. Cohen, and H. H. Hu, *Fluid mechanics* (Elsevier Academic Press, San Diego, 2004). [page 8].

[40] H. T. Rossby, *A study of Bénard convection with and without rotation,* J. Fluid Mech. **36**, 309 (1969). [pages 9, 14, and 15].

[41] G. Zocchi, E. Moses, and A. Libchaber, *Coherent structures in turbulent convection, an experimental study,* Physica A Stat. Mech. Appl. **166**, 387 (1990). [pages 10 and 11].

[42] L. P. Kadanoff, *Turbulent heat flow: structures and scaling,* Phys. Today **54**, 34 (2001). [page 10].

[43] H.-D. Xi, S. Lam, and K.-Q. Xia, *From laminar plumes to organized flows: the onset of large-scale circulation in turbulent thermal convection,* J. Fluid Mech. **503**, 47 (2004). [page 11].

[44] S. Cioni, S. Ciliberto, and J. Sommeria, *Strongly turbulent Rayleigh–Bénard convection in mercury: comparison with results at moderate Prandtl number,* J. Fluid Mech. **335**, 111 (1997). [pages 11 and 14].

[45] C. Sun, H.-D. Xi, and K.-Q. Xia, *Azimuthal symmetry, flow dynamics, and heat transport in turbulent thermal convection in a cylinder with an aspect ratio of 0.5,* Phys. Rev. Lett. **95**, 074502:1 (2005). [pages 19, 21, 78, 79, and 85].

[46] E. Brown and G. Ahlers, *Effect of the Earth's Coriolis force on the large-scale circulation of turbulent Rayleigh-Bénard convection,* Phys. Fluids **18**, 125108:1 (2006). [pages 19, 92, and 96].

[47] E. Brown and G. Ahlers, *Rotations and cessations of the large-scale circulation in turbulent Rayleigh–Bénard convection,* J. Fluid Mech. **568**, 351 (2006). [page 12].

[48] H.-D. Xi, Q. Zhou, and K.-Q. Xia, *Azimuthal motion of the mean wind in turbulent thermal convection,* Physical Review E **73**, 056312:1 (2006).

[49] H.-D. Xi and K.-Q. Xia, *Cessations and reversals of the large-scale circulation in turbulent thermal convection,* Phys. Rev. E **75**, 066307:1 (2007). [pages 11 and 12].

[50] D. Funfschilling and G. Ahlers, *Plume motion and large-scale circulation in a cylindrical Rayleigh-Bénard cell,* Phys. Rev. Lett. **92**, 194502:1 (2004). [page 11].

[51] D. Funfschilling, E. Brown, and G. Ahlers, *Torsional oscillations of the large-scale circulation in turbulent Rayleigh–Bénard convection*, J. Fluid Mech. **607**, 119 (2008). [pages 11, 19, and 92].

[52] H.-D. Xi, S.-Q. Zhou, Q. Zhou, T.-S. Chan, and K.-Q. Xia, *Origin of the temperature oscillation in turbulent thermal convection*, Phys. Rev. Lett. **102**, 044503:1 (2009). [pages 11, 12, and 90].

[53] Q. Zhou, H.-D. Xi, S.-Q. Zhou, C. Sun, and K.-Q. Xia, *Oscillations of the large-scale circulation in turbulent Rayleigh–Bénard convection: the sloshing mode and its relationship with the torsional mode*, J. Fluid Mech. **630**, 367 (2009). [pages 11 and 90].

[54] E. Brown, A. Nikolaenko, and G. Ahlers, *Reorientation of the large-scale circulation in turbulent Rayleigh-Bénard convection*, Phys. Rev. Lett. **95**, 084503:1 (2005). [pages 12 and 17].

[55] E. Villermaux, *Memory-induced low frequency oscillations in closed convection boxes*, Phys. Rev. Lett. **75**, 4618 (1995). [page 12].

[56] E. van Doorn, B. Dhruva, K. R. Sreenivasan, and V. Cassella, *Statistics of wind direction and its increments*, Phys. Fluids **12**, 1529 (2000). [page 12].

[57] R. Verzicco and R. Camussi, *Numerical experiments on strongly turbulent thermal convection in a slender cylindrical cell*, J. Fluid Mech. **477**, 19 (2003). [pages 12, 50, 78, 79, and 86].

[58] H.-D. Xi and K.-Q. Xia, *Flow mode transitions in turbulent thermal convection*, Phys. Fluids **20**, 055104:1 (2008). [pages 12, 19, 94, 95, 96, 97, and 134].

[59] G. Stringano and R. Verzicco, *Mean flow structure in thermal convection in a cylindrical cell of aspect ratio one half*, J. Fluid Mech. **548**, 1 (2006). [page 12].

[60] J. Hart, S. Kittelman, and D. Ohlsen, *Mean flow precession and temperature probability density functions in turbulent rotating convection*, Phys. Fluids **14**, 955 (2002). [page 12].

[61] R. Kunnen, H. Clercx, and B. J. Geurts, *Breakdown of large-scale circulation in turbulent rotating convection*, Europhys. Lett. **84**, 24001:1 (2008). [pages 12, 16, and 21].

[62] J.-Q. Zhong and G. Ahlers, *Heat transport and the large-scale circulation in rotating turbulent Rayleigh–Bénard convection*, J. Fluid Mech. **665**, 300 (2010). [page 12].

[63] S. Weiss and G. Ahlers, *The large-scale flow structure in turbulent rotating Rayleigh–Bénard convection*, J. Fluid Mech. **688**, 461 (2011). [pages 12, 93, 94, 95, 96, 97, 98, and 134].

[64] R. P. J. Kunnen, H. J. H. Clercx, and B. J. Geurts, *Vortex statistics in turbulent rotating convection*, Phys. Rev. E **82**, 036306:1 (2010). [pages 13 and 21].

[65] J. C. R. Hunt, A. A. Wray, and P. Moin, *Eddies, streams, and convergence zones in turbulent flows*, Center for Turbulence Research Report No. CTR-S88 (1988). [pages 13 and 78].

[66] G. Haller, *An objective definition of a vortex*, J. Fluid Mech. **525**, 1 (2005). [page 13].

[67] R. J. Stevens, J.-Q. Zhong, H. J. Clercx, G. Ahlers, and D. Lohse, *Transitions between turbulent states in rotating Rayleigh-Bénard convection*, Phys. Rev. Lett. **103**, 024503 (2009). [page 13].

[68] S. Weiss, R. J. Stevens, J.-Q. Zhong, H. J. Clercx, D. Lohse, and G. Ahlers, *Finite-size effects lead to supercritical bifurcations in turbulent rotating Rayleigh-Bénard convection*, Phys. Rev. Lett. **105**, 224501:1 (2010). [pages 13 and 16].

[69] K. Julien, S. Legg, J. McWilliams, and J. Werne, *Plumes in rotating convection. Part 1. Ensemble statistics and dynamical balances*, J. of Fluid Mech. **391**, 151 (1999). [page 13].

[70] S. Legg, K. Julien, J. McWilliams, and J. Werne, *Vertical transport by convection plumes: modification by rotation*, Phys. Chem. Earth (B) **26**, 259 (2001). [page 13].

[71] R. P. Kunnen, R. J. Stevens, J. Overkamp, C. Sun, G. F. van Heijst, and H. J. Clercx, *The role of Stewartson and Ekman layers in turbulent rotating Rayleigh–Bénard convection*, J. Fluid Mech. **688**, 422 (2011). [pages 13, 110, and 127].

[72] B. Boubnov and G. Golitsyn, *Experimental study of convective structures in rotating fluids*, J. Fluid Mech. **167**, 503 (1986). [page 13].

[73] F. Zhong, R. E. Ecke, and V. Steinberg, *Rotating Rayleigh–Bénard convection: asymmetric modes and vortex states*, J. Fluid Mech. **249**, 135 (1993).

[74] K. Julien, S. Legg, J. McWilliams, and J. Werne, *Rapidly rotating turbulent Rayleigh-Bénard convection*, J. Fluid Mech. **322**, 243 (1996).

[75] R. E. Ecke and Y. Liu, *Traveling-wave and vortex states in rotating Rayleigh–Bénard convection*, Int. J. Eng. Sci. **36**, 1471 (1998). [page 13].

[76] S. Grossmann and D. Lohse, *Scaling in thermal convection: a unifying theory*, J. Fluid Mech. **407**, 27 (2000). [page 14].

[77] S. Grossmann and D. Lohse, *Thermal convection for large Prandtl numbers*, Phys. Rev. Lett. **86**, 3316 (2001).

[78] S. Grossmann and D. Lohse, *Prandtl and Rayleigh number dependence of the Reynolds number in turbulent thermal convection*, Phys. Rev. E **66**, 016305:1 (2002).

[79] S. Grossmann and D. Lohse, *Fluctuations in turbulent Rayleigh–Bénard convection: the role of plumes*, Phys. Fluids **16**, 4462 (2004). [page 14].

[80] R. H. Kraichnan, *Turbulent thermal convection at arbitrary Prandtl number*, Phys. Fluids **5**, 1374 (1962). [page 14].

[81] E. A. Spiegel, *Convection in stars I. Basic Boussinesq Convection*, Annu. Rev. Astron. Astrophys. **9**, 323 (1971). [page 14].

[82] X. Chavanne, F. Chilla, B. Castaing, B. Hebral, B. Chabaud, and J. Chaussy, *Observation of the ultimate regime in Rayleigh-Bénard convection*, Phys. Rev. Lett. **79**, 3648 (1997). [page 14].

[83] X. Chavanne, F. Chilla, B. Chabaud, B. Castaing, and B. Hebral, *Turbulent Rayleigh–Bénard convection in gaseous and liquid He*, Phys. Fluids **13**, 1300 (2001).

[84] J. J. Niemela, L. Skrbek, K. R. Sreenivasan,  and R. J. Donnelly, *Turbulent convection at very high Rayleigh numbers,* Nature **404**, 837 (2000). [page 14].

[85] X. He, D. Funfschilling, H. Nobach, E. Bodenschatz,  and G. Ahlers, *Transition to the ultimate state of turbulent Rayleigh-Bénard convection,* Phys. Rev. Lett. **108**, 024502: (2012). [pages 14 and 18].

[86] J. A. Glazier, T. Segawa, A. Naert,  and M. Sano, *Evidence against ultrahard thermal turbulence at very high Rayleigh numbers,* Nature **398**, 307 (1999). [page 14].

[87] S. Horanyi, L. Krebs,  and U. Müller, *Turbulent Rayleigh–Bénard convection in low Prandtl–number fluids,* Int. J. Heat Mass Transfer **42**, 3983 (1999). [page 14].

[88] G. Ahlers and X. Xu, *Prandtl-number dependence of heat transport in turbulent Rayleigh-Bénard convection,* Phys. Rev. Lett. **86**, 3320 (2001). [page 14].

[89] K.-Q. Xia, S. Lam,  and S.-Q. Zhou, *Heat-flux measurement in high-Prandtl-number turbulent Rayleigh-Bénard convection,* Phys. Rev. Lett. **88**, 064501:1 (2002). [page 14].

[90] R. du Puits, C. Resagk,  and A. Thess, *Breakdown of wind in turbulent thermal convection,* Phys. Rev. E **75**, 016302:1 (2007). [page 15].

[91] C. Sun, Y.-H. Cheung,  and K.-Q. Xia, *Experimental studies of the viscous boundary layer properties in turbulent Rayleigh–Bénard convection,* J. Fluid Mech. **605**, 79 (2008). [page 15].

[92] K.-Q. Xia, C. Sun,  and Y.-H. Cheung, *Large scale velocity structures in turbulent thermal convection with widely varying aspect ratio,* in *Proc. of the 14th International Symposium on Applications of Laser Techniques to Fluid Mechanics* (Lisbon, Portugal, July 7-10, 2008). [pages 15, 79, and 86].

[93] D. Funfschilling, E. Brown, A. Nikolaenko,  and G. Ahlers, *Heat transport by turbulent Rayleigh–Bénard convection in cylindrical samples with aspect ratio one and larger,* J. Fluid Mech. **536**, 145 (2005). [pages 15, 19, and 92].

[94] A. Nikolaenko, E. Brown, D. Funfschilling,  and G. Ahlers, *Heat transport by turbulent Rayleigh–Bénard convection in cylindrical cells with aspect ratio one and less,* J. Fluid Mech. **523**, 251 (2005).

[95] C. Sun, L.-Y. Ren, H. Song,  and K.-Q. Xia, *Heat transport by turbulent Rayleigh–Bénard convection in 1 m diameter cylindrical cells of widely varying aspect ratio,* J. Fluid Mech. **542**, 165 (2005). [pages 15, 19, 21, 54, and 86].

[96] S. Chandrasekhar, *Hydrodynamic and hydromagnetic stability* (Dover, New York, 1981). [page 15].

[97] K. Julien, S. Legg, J. McWilliams,  and J. Werne, *Hard turbulence in rotating Rayleigh-Bénard convection,* Phys. Rev. E **53**, R5557 (1996). [page 15].

[98] P. Vorobieff and R. E. Ecke, *Turbulent rotating convection: an experimental study,* J. Fluid Mech. **458**, 191 (2002). [page 15].

[99] K. Julien, A. Rubio, I. Grooms,  and E. Knobloch, *Statistical and physical balances in low Rossby number Rayleigh–Bénard convection,* Geophys. Astro. Fluid **106**, 392 (2012). [page 16].

[100] D. Nieves, A. M. Rubio, and K. Julien, *Statistical classification of flow morphology in rapidly rotating Rayleigh-Bénard convection,* Phys. Fluids **26**, 086602:1 (2014). [page 16].

[101] R. J. A. M. Stevens, J. Overkamp, D. Lohse, and H. J. H. Clercx, *Effect of aspect ratio on vortex distribution and heat transfer in rotating Rayleigh-Bénard convection,* Phys. Rev. E **84**, 056313:1 (2011). [pages 16, 19, 96, and 134].

[102] J.-Q. Zhong, R. J. A. M. Stevens, H. J. Clercx, R. Verzicco, D. Lohse, and G. Ahlers, *Prandtl-, Rayleigh-, and Rossby-number dependence of heat transport in turbulent rotating Rayleigh-Bénard convection,* Phys. Rev. Lett. **102**, 044502:1 (2009). [page 16].

[103] P. Oresta, G. Stringano, and R. Verzicco, *Transitional regimes and rotation effects in Rayleigh–Bénard convection in a slender cylindrical cell,* Eur. J. Mech. B/Fluids **26**, 1 (2007). [page 16].

[104] R. J. A. M. Stevens, H. J. H. Clercx, and D. Lohse, *Optimal Prandtl number for heat transfer in rotating Rayleigh–Bénard convection,* New J. Phys. **12**, 075005:1 (2010). [page 17].

[105] S. Chaumat, B. Castaing, and F. Chilla, *Rayleigh-Bénard cells: influence of the plates properties Advances in Turbulence IX,* in *Advances in Turbulence IX. Proc. of 9th European Turbulence Conference* (ed. IP Castro & PE Hancock, CIMNe, Barcelona, Spain, 2002). [page 17].

[106] J. Hunt, A. Vrieling, F. Nieuwstadt, and H. Fernando, *The influence of the thermal diffusivity of the lower boundary on eddy motion in convection,* J. Fluid Mech. **491**, 183 (2003). [page 18].

[107] R. Verzicco, *Effects of nonperfect thermal sources in turbulent thermal convection,* Phys. Fluids **16**, 1965 (2004). [page 17].

[108] G. Ahlers, *Effect of sidewall conductance on heat-transport measurements for turbulent Rayleigh-Benard convection,* Phys. Rev. E **63**, 015303:1 (2000). [page 18].

[109] P.-E. Roche, B. Castaing, B. Chabaud, B. Hébral, and J. Sommeria, *Side wall effects in Rayleigh Bénard experiments,* Eur. Phys. J. B **24**, 405 (2001).

[110] R. Verzicco, *Sidewall finite-conductivity effects in confined turbulent thermal convection,* J. Fluid Mech. **473**, 201 (2002). [page 18].

[111] J. J. Niemela and K. R. Sreenivasan, *Confined turbulent convection,* J. Fluid Mech. **481**, 355 (2003). [pages 17 and 18].

[112] R. J. Stevens, D. Lohse, and R. Verzicco, *Sidewall effects in Rayleigh–Bénard convection,* J. Fluid Mech. **741**, 1 (2014). [pages 18 and 49].

[113] F. Chilla, M. Rastello, S. Chaumat, and B. Castaing, *Long relaxation times and tilt sensitivity in Rayleigh Bénard turbulence,* Eur. Phys. J. B **40**, 223 (2004). [page 19].

[114] Y. Gasteuil, W. L. Shew, M. Gibert, F. Chilla, B. Castaing, and J.-F. Pinton, *Lagrangian temperature, velocity, and local heat flux measurement in Rayleigh-Bénard convection,* Phys. Rev. Lett. **99**, 234302 (2007). [page 19].

[115] W. L. Shew, Y. Gasteuil, M. Gibert, P. Metz, and J.-F. Pinton, *Instrumented tracer for Lagrangian measurements in Rayleigh-Bénard convection*, Rev. Sci. Instrum. **78**, 065105 (2007).

[116] R. Ni, S.-D. Huang, and K.-Q. Xia, *Lagrangian acceleration measurements in convective thermal turbulence*, J. Fluid Mech. **692**, 395 (2012). [pages 21, 54, and 59].

[117] O. Liot, A. Gay, J. Salort, M. Bourgoin, and F. Chillà, *Inhomogeneity and Lagrangian unsteadiness in turbulent thermal convection*, Phys. Rev. Fluids **1**, 064406 (2016).

[118] O. Liot, F. Seychelles, F. Zonta, S. Chibbaro, T. Coudarchet, Y. Gasteuil, J.-F. Pinton, J. Salort, and F. Chillà, *Simultaneous temperature and velocity Lagrangian measurements in turbulent thermal convection*, J. Fluid Mech. **794**, 655 (2016). [page 19].

[119] M. Raffel, C. Willert, S. Wereley, and J. Kompenhans, *Particle Image Velocimetry: A Practical Guide*, Experimental Fluid Mechanics (Springer Berlin Heidelberg, 2007). [pages 21, 27, and 36].

[120] R. Adrian and J. Westerweel, *Particle Image Velocimetry*, Cambridge Aerospace Series (Cambridge University Press, 2011). [pages 36 and 37].

[121] J. Westerweel, G. E. Elsinga, and R. J. Adrian, *Particle image velocimetry for complex and turbulent flows*, Annu. Rev. Fluid Mech. **45**, 409 (2013). [page 21].

[122] K.-Q. Xia, C. Sun, and S.-Q. Zhou, *Particle image velocimetry measurement of the velocity field in turbulent thermal convection*, Phys. Rev. E **68**, 066303:1 (2003). [page 21].

[123] C. Sun, K.-Q. Xia, and P. Tong, *Three-dimensional flow structures and dynamics of turbulent thermal convection in a cylindrical cell*, Phys. Rev. E **72**, 026302:1 (2005). [page 78].

[124] C. Sun, Q. Zhou, and K.-Q. Xia, *Cascades of velocity and temperature fluctuations in buoyancy-driven thermal turbulence*, Phys. Rev. Lett. **97**, 144504:1 (2006).

[125] R. Kunnen, B. J. Geurts, and H. Clercx, *Experimental and numerical investigation of turbulent convection in a rotating cylinder*, J. Fluid Mech. **642**, 445 (2010). [pages 21 and 109].

[126] D. Schiepel, D. Schmeling, and C. Wagner, *Simultaneous velocity and temperature measurements in turbulent Rayleigh-Bénard convection based on combined tomo-PIV and PIT*, in *Proc. of 18th International Symposium on Applications of Laser Techniques to Fluid Mechanics* (Lisbon, Portugal, July 4-7, 2016). [page 21].

[127] S. Discetti and F. Coletti, *Volumetric velocimetry for fluid flows*, Meas. Sci. Technol. **29**, 042001 (2018). [page 21].

[128] G. E. Elsinga, F. Scarano, B. Wieneke, and B. W. van Oudheusden, *Tomographic particle image velocimetry*, Exp. Fluids **41**, 933 (2006). [pages 21, 22, 31, and 32].

[129] F. Scarano, *Tomographic PIV: principles and practice,* Meas. Sci. Technol. **24**, 012001:1 (2012). [pages 21, 30, 31, and 33].

[130] B. Wieneke, *Volume self-calibration for 3D particle image velocimetry,* Exp. Fluids **45**, 549 (2008). [pages 22 and 28].

[131] S. Discetti and T. Astarita, *The detrimental effect of increasing the number of cameras on self-calibration for tomographic PIV,* Meas. Sci. Technol. **25**, 084001 (2014). [page 28].

[132] B. Wieneke, *Improvements for volume self-calibration,* Meas. Sci. Technol. (2018), 10.1088/1361-6501/aacd45. [page 22].

[133] J. Westerweel and F. Scarano, *Universal outlier detection for PIV data,* Exp. Fluids **39**, 1096 (2005). [page 23].

[134] A. Melling, *Tracer particles and seeding for particle image velocimetry,* Meas. Sci. Technol. **8**, 1406 (1997). [pages 23 and 25].

[135] T. Chan-Mou, *Mean value and correlation Problems connected with the motion of small particles suspended in a turbulent fluid* (Springer, 2013). [page 23].

[136] T. Koyaguchi, M. A. Hallworth, H. E. Huppert, and R. S. J. Sparks, *Sedimentation of particles from a convecting fluid,* Nature **343**, 447 (1990). [page 25].

[137] M. Okada, C. Kang, K. Oyama, and S. Yano, *Natural convection of a water-fine particle suspension in a rectangular cell heated and cooled from opposing vertical walls: The effect of distribution of particle size,* Heat Transfer—Asian Res. **30**, 636 (2001). [page 25].

[138] K. Gotoh, S. Yamada, and T. Nishimura, *Influence of thermal convection on particle behavior in solid-liquid suspensions,* Adv. Powder Technol. **15**, 499 (2004).

[139] G. Lavorel and M. Le Bars, *Sedimentation of particles in a vigorously convecting fluid,* Phys. Rev. E **80**, 046324:1 (2009).

[140] P. Joshi, H. Rajaei, R. P. Kunnen, and H. J. Clercx, *Effect of particle injection on heat transfer in rotating Rayleigh-Bénard convection,* Phys. Rev. Fluids **1**, 084301 (2016). [pages 25 and 119].

[141] Particle Science Drug Development Services, *Physical Stability of Disperse Systems,* Technical Brief 2009 **1** (2009). [page 25].

[142] R. J. Adrian and C.-S. Yao, *Pulsed laser technique application to liquid and gaseous flows and the scattering power of seed materials,* Appl. Opt. **24**, 44 (1985). [pages 25 and 27].

[143] H. C. Hulst and H. C. van de Hulst, *Light scattering by small particles* (Courier Corporation, 1981). [page 25].

[144] F. Scarano, S. Ghaemi, G. C. A. Caridi, J. Bosbach, U. Dierksheide, and A. Sciacchitano, *On the use of helium-filled soap bubbles for large-scale tomographic PIV in wind tunnel experiments,* Exp. Fluids **56**, 42:1 (2015). [page 26].

[145] A. Schröder, R. Geisler, G. E. Elsinga, F. Scarano, and U. Dierksheide, *Investigation of a turbulent spot and a tripped turbulent boundary layer flow using time-resolved tomographic PIV,* Exp. Fluids **44**, 305 (2008). [page 26].

[146] S. Ghaemi and F. Scarano, *Multi-pass light amplification for tomographic particle image velocimetry applications,* Meas. Sci. Technol. **21**, 127002 (2010). [page 26].

[147] J. Westerweel, *Digital particle image velocimetry — Theory and application,* Ph.D. thesis, Delft University of Technology (1993). [page 27].

[148] J. Westerweel, D. Dabiri, and M. Gharib, *The effect of a discrete window offset on the accuracy of cross-correlation analysis of digital PIV recordings,* Exp. Fluids **23**, 20 (1997). [page 27].

[149] R. Tsai, *A versatile camera calibration technique for high-accuracy 3D machine vision metrology using off-the-shelf TV cameras and lenses,* IEEE J. Robot. Autom. **3**, 323 (1987). [pages 29 and 44].

[150] J. Heikkila and O. Silven, *A four-step camera calibration procedure with implicit image correction,* in *Proc. of IEEE Computer Society Conference on Computer Vision and Pattern Recognition* (IEEE, 1997) pp. 1106–1112. [pages 29, 44, 58, and 65].

[151] S. M. Soloff, R. J. Adrian, and Z.-C. Liu, *Distortion compensation for generalized stereoscopic particle image velocimetry,* Meas. Sci. Technol. **8**, 1441 (1997). [page 29].

[152] R. Gordon and G. T. Herman, *Reconstruction of pictures from their projections,* Commun. ACM **14**, 759 (1971). [page 31].

[153] Y. Censor, *Row-action methods for huge and sparse systems and their applications,* SIAM Rev. **23**, 444 (1981).

[154] D. Mishra, K. Muralidhar, and P. Munshi, *A robust MART algorithm for tomographic applications,* Numer. Heat Tr. B-Fund. **35**, 485 (1999). [page 31].

[155] R. Gordon, R. Bender, and G. T. Herman, *Algebraic reconstruction techniques (ART) for three-dimensional electron microscopy and X-ray photography,* J. Theor. Biol. **29**, 471 (1970). [page 31].

[156] G. T. Herman and A. Lent, *Iterative reconstruction algorithms,* Comput. Biol. Med. **6**, 273 (1976). [page 31].

[157] C. Atkinson and J. Soria, *An efficient simultaneous reconstruction technique for tomographic particle image velocimetry,* Exp. Fluids **47**, 553 (2009). [pages 31, 35, and 46].

[158] L. Thomas, B. Tremblais, and L. David, *Optimization of the volume reconstruction for classical Tomo-PIV algorithms (MART, BIMART and SMART): synthetic and experimental studies,* Meas. Sci. Technol. **25**, 035303:1 (2014). [page 31].

[159] M. Novara, K. J. Batenburg, and F. Scarano, *Motion tracking-enhanced MART for tomographic PIV,* Meas. Sci. Technol. **21**, 035401:1 (2010). [pages 32 and 34].

[160] C. Atkinson, S. Coudert, J. Foucaut, M. Stanislas, and J. Soria, *Thick and Thin Volume Measurements of a Turbulent Boundary Layer using Tomographic Particle Image Velocimetry,* in *Proc. of 8th International Symposium on Particle Image Velocimetry* (Melbourne, Victoria, Australia, August 25-28, 2009) pp. 25–28. [page 32].

[161] G. Elsinga, J. Westerweel, F. Scarano,  and M. Novara, *On the velocity of ghost particles and the bias errors in Tomographic-PIV,* Exp. Fluids **50**, 825 (2011). [page 33].

[162] S. Discetti, *Tomographic particle image velocimetry — Developments and applications to turbulent flows,* Ph.D. thesis, Università degli Studi di Napoli "Federico II" (2013). [page 32].

[163] S. Discetti and T. Astarita, *A fast multi-resolution approach to tomographic PIV,* Exp. Fluids **52**, 765 (2012). [pages 33 and 46].

[164] K. Lynch and F. Scarano, *An efficient and accurate approach to MTE-MART for time-resolved tomographic PIV,* Exp. Fluids **56**, 66:1 (2015). [pages 34 and 46].

[165] B. Wieneke, *Iterative reconstruction of volumetric particle distribution,* Meas. Sci. Technol. **24**, 024008:1 (2012). [pages 34, 35, and 47].

[166] D. Schanz, A. Schröder, S. Gesemann, D. Michaelis,  and B. Wieneke, *Shake the box: a highly efficient and accurate tomographic particle tracking velocimetry (TOMO-PTV) method using prediction of particle positions,* in *Proc. of 10th International Symposium on Particle Image Velocimetry* (Delft, The Netherlands, July 1-3, 2013). [page 35].

[167] D. Schanz, S. Gesemann,  and A. Schröder, *Shake-The-Box: Lagrangian particle tracking at high particle image densities,* Exp. Fluids **57**, 70 (2016). [pages 35 and 47].

[168] C. J. Kähler, T. Astarita, P. P. Vlachos, J. Sakakibara, R. Hain, S. Discetti, R. La Foy,  and C. Cierpka, *Main results of the 4th International PIV Challenge,* Exp. Fluids **57**, 97 (2016). [page 35].

[169] C. E. Willert and M. Gharib, *Digital particle image velocimetry,* Exp. Fluids **10**, 181 (1991). [page 36].

[170] J. Nogueira, A. Lecuona,  and P. Rodriguez, *Local field correction PIV: on the increase of accuracy of digital PIV systems,* Exp. Fluids **27**, 107 (1999). [page 36].

[171] F. Scarano, *Iterative image deformation methods in PIV,* Meas. Sci. Technol. **13**, R1:1 (2002). [page 36].

[172] S. Discetti and T. Astarita, *Fast 3D PIV with direct sparse cross-correlations,* Exp. Fluids **53**, 1437 (2012). [pages 37 and 47].

[173] M. Mendez, M. Raiola, A. Masullo, S. Discetti, A. Ianiro, R. Theunissen,  and J.-M. Buchlin, *POD-based background removal for particle image velocimetry,* Exp. Therm. Fluid Sci. **80**, 181 (2017). [page 46].

[174] F. J. Martins, J.-M. Foucaut, L. Thomas, L. F. Azevedo,  and M. Stanislas, *Volume reconstruction optimization for tomo-PIV algorithms applied to experimental data,* Meas. Sci. Technol. **26**, 085202 (2015). [page 46].

[175] G. Ceglia, S. Discetti, A. Ianiro, D. Michaelis, T. Astarita,  and G. Cardone, *Three-dimensional organization of the flow structure in a non-reactive model aero engine lean burn injection system,* Exp. Therm. Fluid Sci. **52**, 164 (2014). [page 46].

[176] G. Castrillo, G. Cafiero, S. Discetti, and T. Astarita, *Blob-enhanced reconstruction technique*, Meas. Sci. Technol. **27**, 094011 (2016). [pages 46 and 71].

[177] T. Astarita and G. Cardone, *Analysis of interpolation schemes for image deformation methods in PIV*, Exp. Fluids **38**, 233 (2005). [page 47].

[178] T. Astarita, *Analysis of interpolation schemes for image deformation methods in PIV: effect of noise on the accuracy and spatial resolution*, Exp. Fluids **40**, 977 (2006).

[179] T. Astarita, *Analysis of weighting windows for image deformation methods in PIV*, Exp. Fluids **43**, 859 (2007).

[180] T. Astarita, *Analysis of velocity interpolation schemes for image deformation methods in PIV*, Exp. Fluids **45**, 257 (2008). [page 47].

[181] R. Verzicco and P. Orlandi, *A finite-difference scheme for three-dimensional incompressible flows in cylindrical coordinates*, J. Comput. Phys. **123**, 402 (1996). [page 50].

[182] R. Verzicco and R. Camussi, *Prandtl number effects in convective turbulence*, J. Fluid Mech. **383**, 55 (1999). [page 50].

[183] R. J. Stevens, R. Verzicco, and D. Lohse, *Radial boundary layer structure and Nusselt number in Rayleigh–Bénard convection*, J. Fluid Mech. **643**, 495 (2010). [page 50].

[184] O. Shishkina, R. J. Stevens, S. Grossmann, and D. Lohse, *Boundary layer structure in turbulent thermal convection and its consequences for the required numerical resolution*, New J. Phys. **12**, 075022 (2010). [page 50].

[185] B. Hof, C. W. van Doorne, J. Westerweel, and F. T. Nieuwstadt, *Turbulence regeneration in pipe flow at moderate Reynolds numbers*, Phys. Rev. Lett. **95**, 214502 (2005). [page 54].

[186] I. Marusic, B. McKeon, P. Monkewitz, H. Nagib, A. Smits, and K. Sreenivasan, *Wall-bounded turbulent flows at high Reynolds numbers: recent advances and key issues*, Phys. Fluids **22**, 065103 (2010). [page 54].

[187] S. A. Basha and K. R. Gopal, *In-cylinder fluid flow, turbulence and spray models—A review*, Renew. Sust. Energ. Rev. **13**, 1620 (2009). [page 54].

[188] A. K. Agarwal, S. Gadekar, and A. P. Singh, *In-cylinder air-flow characteristics of different intake port geometries using tomographic PIV*, Phys. Fluids **29**, 095104 (2017). [page 54].

[189] S. Tokgoz, G. E. Elsinga, R. Delfos, and J. Westerweel, *Spatial resolution and dissipation rate estimation in Taylor–Couette flow for tomographic PIV*, Exp. Fluids **53**, 561 (2012). [page 54].

[190] S. G. Huisman, D. P. van Gils, S. Grossmann, C. Sun, and D. Lohse, *Ultimate turbulent Taylor-Couette flow*, Phys. Rev. Lett. **108**, 024501 (2012). [page 54].

[191] C. W. H. van Doorne, *Stereoscopic PIV on transition in pipe flow*, Ph.D. thesis, Delft University of Technology (2018). [page 54].

[192] J. Weng, P. Cohen, and M. Herniou, *Camera calibration with distortion models and accuracy evaluation*, IEEE Trans. Pattern Anal. Mach. Intell. , 965 (1992). [page 57].

[193] Y. Tsuji, T. Mizuno, T. Mashiko,  and M. Sano, *Mean wind in convective turbulence of mercury,* Phys. Rev. Lett. **94**, 034501 (2005). [page 78].

[194] V. Kolar, *Vortex identification: New requirements and limitations,* Int. J. Heat Fluid Fl. **28**, 638 (2007). [page 78].

[195] S. Weiss and G. Ahlers, *Turbulent Rayleigh–Bénard convection in a cylindrical container with aspect ratio* $\Gamma = 0.50$ *and Prandtl number* $Pr = 4.38,$ J. Fluid Mech. **676**, 5 (2011). [pages 79, 86, and 87].

[196] G. Berkooz, P. Holmes,  and J. L. Lumley, *The proper orthogonal decomposition in the analysis of turbulent flows,* Annu. Rev. Fluid Mech. **25**, 539 (1993). [pages 84 and 137].

[197] Q. Zhang, Y. Liu,  and S. Wang, *The identification of coherent structures using proper orthogonal decomposition and dynamic mode decomposition,* J. Fluids Structures **49**, 53 (2014). [page 89].

[198] E. Brown and G. Ahlers, *The origin of oscillations of the large-scale circulation of turbulent Rayleigh–Bénard convection,* J. Fluid Mech. **638**, 383 (2009). [page 90].

[199] R. J. Stevens, H. J. Clercx,  and D. Lohse, *Effect of plumes on measuring the large scale circulation in turbulent Rayleigh-Bénard convection,* Phys. Fluids **23**, 095110:1 (2011). [pages 92, 93, and 94].

[200] S. Maurel, J. Borée,  and J. Lumley, *Extended proper orthogonal decomposition: application to jet/vortex interaction,* Flow Turbul. Combust. **67**, 125 (2001). [pages 101, 137, and 141].

[201] J. Borée, *Extended proper orthogonal decomposition: a tool to analyse correlated events in turbulent flows,* Exp. Fluids **35**, 188 (2003). [page 142].

[202] A. Antoranz, A. Ianiro, O. Flores,  and M. García-Villalba, *Extended proper orthogonal decomposition of non-homogeneous thermal fields in a turbulent pipe flow,* Int. J. Heat Mass Transfer **118**, 1264 (2018). [pages 101, 137, and 141].

[203] K. Stewartson, *On almost rigid rotations,* J. Fluid Mech. **3**, 17 (1957). [pages 112 and 121].

[204] K. Stewartson, *On almost rigid rotations. Part 2,* J. Fluid Mech. **26**, 131 (1966).

[205] D. W. Moore and P. G. Saffman, *The structure of free vertical shear layers in a rotating fluid and the motion produced by a slowly rising body,* Phil. Trans. R. Soc. Lond. A **264**, 597 (1969).

[206] G. Van Heijst, *The shear-layer structure in a rotating fluid near a differentially rotating sidewall,* J. Fluid Mech. **130**, 1 (1983).

[207] G. Van Heijst, *Source-sink flow in a rotating cylinder,* J. Eng. Math. **18**, 247 (1984). [pages 112 and 121].

[208] J. Hart and D. Ohlsen, *On the thermal offset in turbulent rotating convection,* Phys. Fluids **11**, 2101 (1999). [page 128].

[209] B. Boubnov and G. Golitsyn, *Temperature and velocity field regimes of convective motions in a rotating plane fluid layer,* J. Fluid Mech. **219**, 215 (1990). [page 128].

[210] H. J. Fernando, R.-R. Chen,  and D. L. Boyer, *Effects of rotation on convective turbulence,* J. Fluid Mech. **228**, 513 (1991). [page 128].

[211] S. Horn and J. M. Aurnou, *Regimes of Coriolis-Centrifugal Convection,* Phys. Rev. Lett. **120**, 204502 (2018). [page 128].

[212] J. L. Lumley, *The structure of inhomogeneous turbulent flows,* Atmospheric Turbulence and Radio Wave Propagation; ed. A.M. Yaglom, V.I. Tatarski, Moscow, *1967,* , 166 (1967). [page 137].

[213] D. D. Kosambi, *Statistics in function space,* J. Indian Math. Soc. **7**, 76 (1943). [page 137].

[214] M. Loève, *Functions aleatoire de second ordre,* Revue Science **84**, 195 (1946).

[215] K. Karhunen, *Zur spektraltheorie stochastischer prozesse,* Ann. Acad. Sci. Fennicae Ser. AI **34** (1946).

[216] V. S. Pugachev, *The general theory of correlation of random functions,* Izv. Akad. Nauk. Ser. Mat. **17**, 401 (1953).

[217] A. Obukhov, *Statistical description of continuous fields,* Tr. Geophys. Int. Acad. Nauk. USSR **24**, 3 (1954). [page 137].

[218] P. Holmes, J. L. Lumley, G. Berkooz, and C. W. Rowley, *Turbulence, coherent structures, dynamical systems and symmetry* (Cambridge University Press, 2012). [pages 137 and 139].

[219] J. N. Kutz, *Data-driven modeling & scientific computation: methods for complex systems & big data* (Oxford University Press, 2013).

[220] S. Wolkwein, *Proper orthogonal decomposition: theory and reduced-order modelling,* Lecture notes (August 27, 2013), Department of Mathematics and Statistics, University of Konstanz, Germany (2013). [page 137].

[221] L. Sirovich, *Turbulence and the dynamics of coherent structures. I. Coherent structures,* Quart. Appl. Math. **45**, 561 (1987). [page 141].

# List of publications

## Journal articles

1. **G. Paolillo**, C.S. Greco and G. Cardone, *Novel Quadruple Synthetic Jet Device: Flow Field and Acoustic Behavior*, AIAA J., **55**, 2241 (2017).

2. **G. Paolillo**, C. S. Greco and G. Cardone, *The evolution of quadruple synthetic jets*, Exp. Therm. Fluid Sci., **89**, 259 (2017).

3. C. S. Greco, **G. Paolillo**, A. Ianiro, G. Cardone and L. De Luca, *Effects of the stroke length and nozzle-to-plate distance on synthetic jet impingement heat transfer*, Int. J. Heat Mass Transf. **117**, 1019 (2018).

4. C. S. Greco, **G. Paolillo**, M. Contino, C. Caramiello, M. Foggia and G. Cardone, *3D temperature mapping of a ceramic shell mold in investment casting process via Infrared Thermography*, IEEE Trans. Industr. Inform., under review.

5. **G. Paolillo**, C. S. Greco and G. Cardone, *Impingement heat transfer from quadruple synthetic jets*, Int. J. Heat Mass Transf., under review.

6. C.S. Greco, **G. Paolillo**, T. Astarita and G. Cardone, *Synthetic jet-based flow control of the von Kármán street behind a circular cylinder*, in preparation.

## Conference contributions

1. **G. Paolillo**, C. S. Greco and G. Cardone, *Characterization of a quadruple synthetic jet device*, In Proc. of 10th Pacific Symposium on Flow Visualization and Image Processing (PSFVIP-10), June 15-18, 2015, Naples, Italy (2015).

2. G. Castrillo, **G. Paolillo**, M. Contino, G. Cafiero and T. Astarita *Flow field features of fractal jets*, In Proc. of 18th International Symposium on the Application of Laser and Imaging Techniques to Fluid Mechanics (LXLASER2016), July 4-7, 2016, Lisbon, Portugal (2016).

3. C. S. Greco, **G. Paolillo**, C. Caramiello, M. Di Foggia and G. Cardone, *3D temperature map reconstruction of a ceramic shell mold in investment casting process*, Quantitative InfraRed Thermography (QIRT 2016), Gdańsk University of Technology, July 4-8, 2016, Poland (2016).

4. **G. Paolillo**, C. S. Greco and G. Cardone, *Time-averaged heat transfer in confined impinging quadruple synthetic jets*, 9th World Conference on Experimental Heat Transfer, Fluid Mechanics and Thermodynamics (ExHFT-9), June 11-15, 2017, Igazu Falls, Brazil (2017).

5. C. S. Greco, **G. Paolillo**, M. Contino, C. Caramiello, M. Di Foggia and G. Cardone, *Investment casting process: 3D temperature map reconstruction of a ceramic shell mold*, 9th World Conference on Experimental Heat Transfer, Fluid Mechanics and Thermodynamics (ExHFT-9), June 11-15, 2017, Igazu Falls, Brazil (2017).

6. C. S. Greco, G. Pascarella, **G. Paolillo**, T. Astarita and G. Cardone, *Synthetic jet-based control of a cylinder wake*, Italian Association of Aeronautics and Astronautics XXIV International Conference (AIDAA 2017), September 18-22, 2017, Palermo-Enna, Italy (2017).

7. **G. Paolillo**, C. S. Greco, G. Pascarella, T. Astarita and G. Cardone, *Control of von Kármán vortex shedding through synthetic jet technology: momentum coefficient and non-dimensional frequency effects*, 14th International Conference on Fluid Control, Measurements and Visualization (FLUCOME 2017), October 8-12, 2017, Notre Dame, Indiana, USA (2017).

8. C. S. Greco, **G. Paolillo** and G. Cardone, *Thermo-fluid-dynamic analysis of innovative synthetic jet devices*, IOP Conf. Ser. Mater. Sci. Eng., **249**, 012001 (2017).

9. **G. Paolillo**, C. S. Greco, T. Astarita and G. Cardone, *Tomographic particle image velocimetry of Rayleigh-Bénard convection in a cylindrical sample*, International Conference on Rayleigh-Bénard Turbulence, May 14-18, 2018, Enschede, The Netherlands (2018).

10. **G. Paolillo**, C. S. Greco, T. Astarita and G. Cardone, *Three-dimensional velocity measurements of Rayleigh-Bénard convection in a cylinder*, In Proc. of 18th International Symposium on Flow Visualization (ISFV-18), June 26-29, 2018, Zurich, Switzerland (2018).

11. C. S. Greco, **G. Paolillo**, T. Astarita and G. Cardone, *Flow control of a cylinder wake using synthetic jet technology*, 19th International Symposium on the Application of Laser and Imaging Techniques to Fluid Mechanics (LXLASER2018), July 16-19, 2018, Lisbon, Portugal (2018).

12. **G. Paolillo** and T. Astarita, *A novel camera model for calibrating optical systems including cylindrical windows*, In Proc. of AIAA Science and Technology Forum and Exposition (AIAA SciTech 2019), January 7-11, 2019, San Diego, California (2019).

# Acknowledgements

On my train home, looking back to the last three years, my Ph.D. seems to have lasted no more than one of my daily travels to and back from Napoli. Considering that I usually travel (and now I am indeed travelling) with Trenitalia, this is saying a lot! Every past event has slipped away through these years... almost every, if I think how long it used to take to carry out a single set of measurements on thermal convection! Anyway, I could not help but notice how much I have grown and how much I have learnt facing all the challenges of this tough, but wonderful, time.

Definitely, I am sure that the best of my luck has been to meet and interact with great people, each of whom, in a direct or incidental way, has left their mark on me and on my education.

The first person I would like to thank is my supervisor prof. Gennaro Cardone. For the opportunity to undertake this path, for the continuous support you provided me, for your teachings, I will never stop saying thank you, Gennaro. I think that often you trust me more than I deserve and, although this puts some pressure on me, I really appreciate it. Your enthusiasm in designing and assembling experimental apparatuses, making equipment work properly, is one of the most straightforward examples of passion for our work to me. Every time I thought I had explored all the possibilities on both practical and theoretical sides, your empirical solutions showed me that there is always one more feasible way you have not conceived or thoroughly considered.

The second person in the list of people worthy of acknowledgments is Carlo (Greco). Actually, I believe that there are very few things I could say that you do not know. Your ability to bear with me in every circumstance is second only to my ability to tease you in the same circumstances. I am not the only one convinced that you deserve a sort of award or a statue for your infinite patience with me and my uncountable scientific (and sociological) problems. Thank you for everything from your side and apologizes for all the pain I gave you every now and then. Not only a friend, you have been also a reference point for me and I admit to envy your insight into physics and other various disciplines, including art and medicine with your predilection for Kandinskij and heart, but excluding math and computer science. I also admire your experimentalist skills, sufficient to obviate any deficiency in any experimental setup (including my presence nearby).

Very special and affectionate thanks go to Tommaso (prof. Astarita). You have been in fact my third tutor, constantly ready for a phone or Skype call. 40% is not a conservative estimate of the percentage of my Ph.D. working time spent talking with you. There has been no conversation of ours, in which I did not learn something unknown and relevant. I soon realized that it is practically impossible to keep pace with your frequency of thinking and formulating a correct perspective or solution of the problem under investigation. My strategy is simply operating at a sub-harmonic (very small) frequency so as to correlate most of your encyclopedic information with my developing ideas.

These years were made so enjoyable also by the presence of great lab colleagues. Contino (Mattia), we have shared all the kinds of misadventure that a respectable Ph.D. candidate should face. Any attempt from my side to summarize them would be incomplete and sacrilegious, so let me overlook and just thank you for having endured until the very end. At least until today... I am sure I will always remember the memes that made our darkest days so easy and pleasant (that cow is still waiting for our company!). I wish the best for your future, Mattia, and I hope to stay in contact. Massi (Cuono Massimo Crispo), although you did use minimal words to thank me in the acknowledgements section of your Ph.D. thesis, I cannot avoid commemorating all our great moments, which prof. Greco and everyone in the lab can witness. I miss your afternoon snack breaks when you used to visit our office with the best intentions. What a gaping hole have you left after your departure! I will never forget your scrupulous care and the tape used to fix every cable and unidentified moving object during your experiments. Thank you for keeping me company on phone often when I was in the Netherlands. Giusy (Castrillo), you were my gossip friend in the lab environment. I remember our long conversations, during the working hours and afterward, walking to the train station. Thank you for those pleasant moments and for all your wise advices on several matters.

Many thanks to Jack (Gioacchino Cafiero), who shared a lot of his experience and knowledge with me in the very early phase of my Ph.D. career and proved always available also in the following, and to Simone (Boccardi), who was always so entertaining and often obviated my late departures from the lab with a ride in his car or scooter. I would like also to mention Umberto (Sabella), who joined our group in the last year and often teaches me about culture and traditions in Napoli, and about the differences between "friarielli" and "broccoli". And above all: Umberto, thank you for all the coffees prepared in the middle of the morning and after lunch!

I am very happy to thank Peppe (Giuseppe Sicardi), our *super* technician. The second award/statue for patience with me is due to you. You have solved a countless number of diplomatic, bureaucratic and administrative issues associated with or caused by me, apart from the ordinary technical problems. You have also won as many unfair bets I did never take indeed. However, all the sfogliatellas I have offered up to now cannot repay your immense help with a wide variety of hitches. Thank you very, very much!

I must also mention all the people who joined my research group in the past and had a role in my Ph.D. career through both stimulating discussions and enjoyable moments spent during their visits to the lab: Francesco Avallone, Giuseppe Ceglia, Andrea Ianiro and Stefano Discetti. I would like to thank prof. Stefano Discetti also for the meticulous review of my Ph.D. thesis and for the thought-provoking comments and suggestions on related present and future work. My gratitude also to the colleagues of Piazzale Tecchio, in particular Enrico Maria De Angelis, Matteo Chiatto and Andrea Palumbo.

Heartfelt thanks to prof. Giovanni Maria Carlomagno for the lovely conversations over these years and his precious suggestions in matter of design of experimental setups and data analysis. I have been always fascinated by his tales on how the university and research used to be in his day, valuable life lessons to ponder on.

My stay in the Netherlands and the period of study at the University of Twente were one of the best experiences of the last three years. Thus, I would like to gratefully thank prof. Roberto Verzicco for giving me the opportunity to join the Physics of Fluids group and undertake such experience. In Enschede, I have found very nice people operating in an international and stimulating environment. Among them, I must thank Joanita Leferink for her constant support with practically any problem and Dr. Shen Chong, who helped me with setup of simulations in the early phase of my numerical activity. Many thanks to Dr. Richard Stevens, who was always ready to solve any issue I experienced with the simulations and provide useful suggestions for improving my work. I am also very grateful to him for reviewing the present dissertation. The numerical simulations reported in this work were performed with the SURFSara Cartesius supercomputer located in Amsterdam. I gratefully acknowledge the PoF group for giving me access to this facility and total autonomy in performing the numerical simulations.

Beyond the university sphere, my friends have constantly stood beside me in this journey. Friendship is a dynamic process, as chaotic as turbulent thermal convection. Although I have never carried out a systematic investigation of the former as of the latter, I have understood that the fundamental point is: no matter how far from each other the bulk flow of our lives takes us, my friends, soon or later we shall meet again and stick together; what we can do in the meanwhile, in the intermediate twists and turns of such a process, is to preserve our reciprocal love. This is my wish for us in the future. Thank you all: Camillo, Flavia, Floriana, Elena, Annalucia, Stefania, Valeria, Andrea, Peppe, Clarissa, Rossella, Alessandro, Nicoletta, Gerardo, Maria Teresa, Annamaria, Alfonso, Emilio. I would also like to thank those friends who, despite the distances and their personal business, are still present in my days: Sara, Claudia, Otello and Roberta.

*Dulcis in fundo*, thanks to my family. Mom and dad, all my achievements would not be possible without your encouragement, your sacrifices and your love. I know that often I do not look grateful to you, but if I work hard, if I do my best, if I never give up, I do it for you, for your pride and faith in me are the greatest rewards. I dedicate this work to you, as well as you devote your days to me and Marilena. Marilena, I could not ask for a better sister. I have always been inspired by your advices and your opinions often change my mind and lead me towards the right choices. I wish you and Valentino all the joys that a married life can offer.

Finally, my thoughts are with my grandparents. Today, I can still enjoy only the hugs and kisses of nonna Lilina, but I know as well, I strongly feel that, right now, nonno Gerardo, nonna Maria and nonno Salvatore are here by my side, in this train, in this journey. And this belief is one of my main sources of strength.

Arriving in Pagani, December 6th, 2018,<br>
Gerardo Paolillo

Youcanprint
Finito di stampare nel mese di marzo 2019